Astronomical Enigmas

Astronomical Enigmas

Life on Mars, the Star of Bethlehem & Other Milky Way Mysteries

MARK KIDGER

THE JOHNS HOPKINS UNIVERSITY PRESS
Baltimore & London

© 2005 The Johns Hopkins University Press
All rights reserved. Published 2005
Printed in the United States of America on acid-free paper
9 8 7 6 5 4 3 2 I

The Johns Hopkins University Press
2715 North Charles Street
Baltimore, Maryland 21218-4363
www.press.jhu.edu

Library of Congress Cataloging-in-Publication Data

Kidger, Mark R. (Mark Richard), 1960–
    Astronomical enigmas : life on Mars, the Star of Bethlehem, and other
Milky Way mysteries / Mark Kidger.
        p.   cm.
    Includes bibliographical references and index.
    ISBN 0-8018-8026-2 (acid-free paper)
    I. Astronomy—Popular works.   I. Title.
    QB43.3.K54 2004
    520—dc22   2004008937

A catalog record for this book is available from the British Library.

Page 298 constitutes an extension of this copyright page.

Contents

Color galleries follow pages 74 and 170

Why Study the Universe?

The human race spends millions of dollars each year on scientific research. Much of this research is what scientists call "applied research"—that is, research directed toward solving a specific problem: developing an AIDS vaccine, using satellite images to discover vital resources such as fresh water and valuable minerals or to locate diseased crops, studying the atmosphere in order to improve weather forecasting, building faster, safer and more energy efficient means of transport, searching for asteroids that may one day be a threat to our planet, and so on. But an important part of all research deals with highly theoretical issues. Did the universe start with the Big Bang, or has it always existed? Will the universe go on expanding forever? What happens in a supernova? What is a star like inside? How was the Moon formed? Where do comets come from? What's inside an atom?

Such questions, while certainly interesting, can seem a matter of intellectual indulgence. Why devote so much time and effort to apparently unproductive research? Why build ever larger telescopes to study distant galaxies or spend countless sums of money on something as "useless" as exploring the planets? Part of the answer, of course, lies in the simple joy of discovery. Then again, one can never be sure what the consequences of seemingly abstract research will be. Whether these consequences are visible in the short term or the long term, they often extend far beyond anything the original researchers might have imagined.

Early in the nineteenth century an English bookbinder named Michael Faraday became interested in electricity. In 1821 he succeeded in building the first dynamo, an invention that would one day make the large-scale generation of electricity possible. At the time, however, with the Industrial Revolution just getting underway, it is unlikely that anyone could have foreseen such a development—although, reputedly, when Faraday was asked by a skeptical prime minister what use electricity was, he replied with remarkable prescience: "Rest assured, one day you will tax it." To take another example: in the early 1950s two researchers invented a process they called "light amplification by stimulated emission of radiation," better known today as the laser. When a journalist inquired what concrete applications their discovery might have, they reportedly answered: "None that we can imagine." In fact, some years later, Arthur Schawlow, who worked with Charles Townes at Bell Labs on the development of the laser, recalled of the invention: "We thought it might have some communications and scientific uses, but we had no application in mind. If we had, it might have hampered us."[1] Such is typical of so-called pure research—and it is equally typical that a facility such as Bell Labs should support research that, at the time, must have appeared to have little practical interest.

Needless to say, fifty years later we are surrounded by devices that use lasers—CD players and computer printers, supermarket bar-code readers, even blackboard pointers. Lasers are used to build secure communications networks and in surgery to carry out bloodless, precision cutting, as well as in countless other commercial and industrial processes. This is one reason why so many universities and research facilities have commercial applications departments: the electronics employed in today's 10-meter telescope will turn up in your television tomorrow and in your car the day after. The technology of the space program is no exception. The medical instruments originally designed to monitor the health of the Apollo astronauts have been an integral part of intensive care equipment in hospitals since the mid-1970s. The personal computers in our homes and workplaces are direct descendants of the computers that were developed to control manned spaceflights in the 1960s. In short, almost all research pays off in the end, and even highly theoretical research tends to pay off in unexpected ways.

But there is a second reason that we seek answers to abstract questions—and that is the real purpose of this book. While it is possible to justify pure research on the grounds that it can have practical benefits, and it is important to be aware of these benefits, I would not want the reader to think exclusively in such terms. Think instead of a more profound reason for investigating the cosmos, for asking questions that may appear odd and irrelevant to our workaday lives: the quest for knowledge. Millions of years of evolution have endowed us with an innate curiosity and with the desire to find solutions to puzzles. Columbus felt obliged to carry out a crazy experiment—to demonstrate that the Earth is round by sailing west to India. Stephen Hawking feels the urge to investigate the interior of a black hole even though we cannot even observe black holes directly (at least not yet). Sir George Mallory wanted to climb Everest "because it is there," and astronomers likewise feel compelled to explore the most distant galaxies.

In a 1976 BBC documentary about particle physics, Nobel Prize–winning physicist Richard Feynman explained that he was once asked by a group of congressional visitors what Fermilab—the Fermi National Accelerator Laboratory, in Illinois, home of some of the greatest particle accelerators on Earth—contributed to national defense. "Nothing," he replied. "It just makes the country worth defending." In the same documentary, Hawking was asked to justify why we pursue abstruse theories about the universe. He pointed to his children and said, "Children always ask questions. Why this and why that. It's what differentiates us from the apes." Hawking hit the nail on the head. The human race has evolved because it has constantly asked questions and set itself challenges. If we had not, we would probably still be living in caves, or possibly be extinct—just another failed experiment of nature, a species that did not make the grade. Just think of the man-apes in Stanley Kubrick's *2001: A Space Odyssey*. They began to flourish only when one of them realized that a bone could be used as a weapon to increase the reach and strength of his blow and thereby enable him to hunt more effectively for food.

This book is not a justification of theoretical science, nor is it intended to be one. But it does illustrate our need to pose and answer ever more complicated questions. Our universe is a fascinating place, full of grandeur and surprises—and the more we learn about the universe, the better we will understand our place in it. One way to appreciate the fragility of our Earth is to compare it with other planets, for in so doing we will see what our planet could have become (and might still). Our solar system offers a wealth of knowledge to the human race, but it also contains some urgent warnings. Although some of the questions we ask may seem of little relevance to a world full or poverty and suffering, we can never be sure. It may be that by asking them we have helped to ensure the survival of the human race.

The eclectic list of questions in this book has been compiled from a variety of sources: university students, amateur astronomers, and interested members of the public. The questions have been gathered into three sections. The first of these looks at the sky from the perspective of our forebears. Starting from what may be the most ancient astronomical observatory in the world, we will move on to examine how the stars were named and the heavens put in order and then to explore what is probably the most mysterious star ever to tax the ingenuity of astronomers. To close our four-thousand-year odyssey we will introduce ourselves to one of the Earth's most famous and distinguished—and reliable—visitors.

In the second section we will take up three questions concerning the solar system that have at times provoked heated argument among astronomers. The subject of the Moon's origin has sparked a considerable amount of debate, although nothing like the passions raised by the question of whether there is, or was, life on Mars. In turn, that controversy pales in comparison to the huge international furor produced by a small, frozen ice ball at the edge of the solar system. To close this section we will learn how astronomers are able to deduce things about places that no human being has ever visited and quite possibly never will. When, for example, astronomers tell us that the surface of Pluto is so cold that methane freezes on it, just how do they know this? By way of an answer we will discover what astronomy's invisible eyes can tell us about volcanoes on one of Jupiter's moons, about what lies beneath the dense clouds that cover the surface of Venus, and about what the nucleus of a comet is made of.

In the final section, we will look into the future and consider two potential disasters that could spell the end of our existence. One is the threat that the Earth could be struck by an asteroid; the other is the possibility that the greenhouse effect could someday do to us what it has done to the two planets closest and most similar to our own. Of course, even supposing that we survive such dangers, we could very well bring about our own end simply by exhausting the resources of

our planet. There is a solution staring us in the face, however, provided we are willing to step into space to secure our tomorrows. To think that human beings have no business anywhere other than on the Earth is at best shortsighted, and quite conceivably a catastrophic mistake. To finish, then, we will come full circle and look at our own origins. How did we, and our planet, get here in the first place? The answer, we will discover, is that we are truly children of the stars.

Some of the questions posed in this book may seem to you important and very much relevant to the human race; others may strike you as hopelessly abstract, or possibly even trivial. One way or another, though, I hope that you will find something in the whole that is thought provoking or enlightening—or just plain fun.

PART I

What Our Forebears Knew

IT IS NOT NEWS THAT IN MANY BRANCHES OF SCIENCE our knowledge and technology have advanced so rapidly that research done twenty years ago may well have been superseded by new studies carried out with improved instruments and techniques and grounded in more sophisticated theories. At the cutting edge of science (including astronomy), work only five years old, or even just a few months old, may be totally outdated. We are, moreover, only human, and one very human trait is to assume that those who went before us did not understand things as well as we do. We thus tend to view our most distant forebears as little better than ignorant savages, although to do so is to sell them short. More than four thousand years ago our Stone Age predecessors were observing the sky, and they were probably far more knowledgeable than we generally suspect. Certainly by 500 B.C. our ancestors had mapped the heavens and named the stars and had populated the sky with the heroes, epic adventures, and animals of all kinds still familiar to us—and the very first star mappers that we are aware of date back to at least a thousand years earlier. Long before that, some of them may have figured out how to use carefully aligned stones to track the movements of the Sun and Moon.

In the first section of this book we will explore a little of the remote and not quite so remote history of astronomy. We will also investigate two famous astronomical mysteries and then meet an old friend. The story will take us around the world, from a cold and windy plain in southern England, to China, Babylon, Palestine, Greece, central Europe, and Korea. On the way we will discover that we sometimes have little idea of what our ancestors actually knew, but equally we will discover that they knew much more than we give them credit for.

CHAPTER I

Stonehenge: Monument or Megalithic Observatory?

High on the itinerary of most visitors to England is a ring of huge stones that stands on the vast expanse of Salisbury Plain, some 15 kilometers outside the town of Winchester—just about halfway between the village near Bristol where I grew up and the city of Southampton. Tourists will either be stunned or disappointed by their visit.

Stonehenge is just one of perhaps a thousand ancient circles of stone and freestanding megaliths scattered throughout the British Isles and northern France, the lands of the ancient Celts. The stone circles range in size from the one at Keel Cross, in County Cork, Ireland, which has a diameter of only 2.75 meters, to the immense circle at Avebury, about 20 kilometers north of Stonehenge, with a diameter of 340 meters. An extensive mythology has developed around these sites, linking them with, among other things, druids and pagan rites and with mystical powers. Despite overwhelming archaeological evidence to the contrary, some people still hold to the legendary belief that King Arthur built Stonehenge, with or without the help of the wizard Merlin. Sites similar to Stonehenge, such as Avebury and Glastonbury, in England, and Carnac, in France, have attracted equally fanciful legends. Glastonbury and its spectacular Tor—a hill outside the town—are rightly famous, but, surprisingly, Avebury is relatively little known and visited far less often than Stonehenge. Perhaps this is because Stonehenge is a more compact site in an isolated location, such that its towering stones dominate the landscape, whereas Avebury is so large that part of a village lies within its circle. Stonehenge imposes itself on the visitor, but the grandeur of the structure at Avebury can only be fully appreciated from the air.

Unfortunately, in our fascination with Stonehenge, we have not always been respectful of the site. As early as 1900, two of the upright stones had fallen owing to the depredations of visitors. The site's then owner, Sir Edward Antrobus, a wealthy local landowner, fenced in the area and began charging an entry fee.[1] In 1915, at the height of the First World War, Sir Edward sold the site to another local landowner, Sir Cecil Chubb, and in 1918 Stonehenge was presented to the nation by its new owner. Sir Edward's efforts to protect the site notwithstanding, by 1979 it had become obvious that some of the stones were suffering considerable damage, once again thanks to the activities of tourists. Since then visitors have not been allowed inside the circle, and security guards are on duty twenty-four hours a day to protect the stones from vandals.

After a rather long ride by car from London, or else by train followed by a bus or an expensive taxi ride from Winchester, one somehow expects more from what is arguably the most famous ancient monument in the world after the Pyramids. Sadly, it is easier to buy a hot dog at Stonehenge than to learn in any detail how and when the circle was built. Evidently the well-known English tradition of understatement applies even to our most famous attractions.[2] Although extensive improvements are promised by English Heritage, the organization responsible for running Stonehenge, and a plan for the site's restoration was published in 2003, the present-day visitor will discover little hint of the many legends that surround these famous stones. Nor is there currently a visitors' center where one might find displays showing how the stones were positioned, or a site plan illustrating the phases of construction, or exhibits of relics from archaeological digs, or even someone to answer questions. English Heritage runs a small souvenir shop, and a host of unofficial souvenir stands line the entranceway. As one crosses the field toward the fence that surrounds the stones themselves, one encounters a short series of panels intended to illustrate a "walk back in time." Astonishingly, a major roadway passes close by the monument. In fact, it crosses one of the ancient earthworks—which most visitors never even realize exist—that form part of the extended Stonehenge site. The famous Heel Stone stands right alongside the road, protected from the stream of speeding cars and trucks by no more than a length of chain-link fence and just a few—a very few—meters of distance. English Heritage's long-awaited plan for the restoration of Stonehenge includes a project to bury part of this road in a tunnel, although critics argue that the proposed tunnel is far too short and that the road should instead be kept completely out of sight. Despite its unprepossessing surroundings, however—to say nothing of the bitter wind that whips across Salisbury Plain in winter—nothing can ultimately detract from the majesty of Stonehenge.

This mammoth stone structure remains in many ways a complete mystery. Why was it built? How and why were some of its stones brought all the way from southwestern Wales? Why was it constructed in three stages, over a period of more than a millennium? How was it used? Was it the site of sacred rites and bloody sacrifices performed by the druids? Was it a temple, or was it an example of ancient "earth magic"—the mystical powers that supposedly reside in certain key sites, points, and lines that cross the British countryside? Or was Stonehenge an astronomical observatory of sorts, very possibly the first ever to be built? These questions will probably never be answered. But then perhaps, as with so many mysteries, Stonehenge is all the more enticing and romantic precisely because there are so many things we will never know with certainty.

The Broken Circle

I was lucky enough to make a visit to Stonehenge in the early 1990s in connection with a project—sadly, later abandoned—to build a half-scale reproduction of the site at El Museo de las Ciencias y el Cósmos in La Laguna, the town on the island of Tenerife (the largest of the Canary Islands) where I work. An artist and I traveled to the site to study it and take photographs, and to attempt to figure out how to produce a realistic reproduction of stone using fiberglass. All this entailed obtaining special permission from English Heritage to enter the circle of stones. Although permission was readily granted, it was only with very exacting conditions. The visit was to be made in December, when Stonehenge closes to visitors at 4:00 P.M. We would be allowed to enter the circle only after the last visitor had left and, even then, only under supervision. Worse still, we would have to leave by 5:00, by which point it would be almost completely dark: in southern England at that time of year the Sun sets well before four in the afternoon, and a long twilight follows. It had been a grey, cold, overcast day, and an icy wind was sweeping off the plain. The light was already quickly fading when we were finally escorted in.

Tourists are permitted only to walk around the site some twenty or thirty meters away from the stones themselves, along a fixed route that does not even include the Heel Stone. From such a distance you can appreciate that the central stones are big, but you almost certainly do not appreciate *how* big. It is not until you stand beside them that you realize how truly enormous they are. To lift such stones by sheer muscle power must have been a task of nearly unimaginable magnitude. Even the relatively small bluestones (so called because they take on a bluish color when wet) weigh some 4 to 5 tonnes each—the sort of load that normally calls for quite a respectable crane.[3] To put it another way, in order to lift one of the bluestones one would need to round up perhaps fifteen Olympic Gold Medal superheavyweight champions. But a bluestone is a trifling matter compared to the estimated 45-tonne weight of the huge sarsen stones.

As has been known for some time, one of the most remarkable features of Stonehenge is the fact that, on the day of the summer solstice the Sun, when viewed from the center of the circle, will be seen to rise directly behind the Heel Stone. The summer solstice is thus the occasion for some interesting festivities at Stonehenge. Roughly three centuries ago the writer and antiquarian John Aubrey suggested that Stonehenge had been built as a temple by the druids, the priests of an ancient Celtic cult. Many people who style themselves as modern-day druids thus travel to the site to celebrate the solstice in the manner of their alleged forebears. This entirely innocent ceremony has on occasion created havoc, as the Wiltshire constabulary attempts to prevent the hundreds or thousands of other people who flock to Stonehenge on that day—some of whom are not as orderly and polite as the modern-day druids—from gaining access to the rather small area around the inner circle of stones.

John Aubrey was, in fact, completely wrong. He associated the construction of Stonehenge with the Celtic people who inhabited Britain at the time of the Roman invasion in 55 B.C. What Aubrey did not know, and could not have known at the time, was that Stonehenge was already upwards of two thousand years old when Julius Caesar ruled and may well have been a ruin even then. Wider still of the mark are those who link the site to King Arthur. According to Arthurian legend, the stones were brought from Ireland in the fifth century to serve as part of a memorial to three hundred British noblemen who had been slaughtered by a treacherous Saxon leader. When the not inconsiderable physical strength of Arthur and his followers proved insufficient to move the stones across the sea to England, Merlin stepped in and moved them by magic. But once again the timing is hopelessly wrong: the construction of Stonehenge began some thirty-five hundred years before King Arthur lived (if indeed he ever did). Whatever mystical abilities Merlin may have possessed, the capacity for time travel is not one habitually attributed to him.[4]

Before we go further, then, we must look at how and by whom Stonehenge was built, and when.

The Construction of Stonehenge

Part of the legend of Stonehenge consists of often wild speculations about who built it. They were not, as we have seen, the Celtic druids, who came to the British Isles many centuries after Stonehenge was constructed. Nor, of course, were they in any sense the "cavemen" conjured up by television's famous Flintstones or by Raquel Welch in the film *One Million Years B.C.* The best guess seems to be that the site was originally created by people of the late Neolithic period, otherwise known as the late Stone Age. These were probably descendants of the original settlers of the British Isles, who crossed over the land bridge from the European continent before it disappeared around ten thousand years ago, allowing the waters of the Atlantic and the North Sea to mingle in the newly formed English Channel. No matter who first began work on the site, however, its construction was completed by the Beaker people. So named for their distinctively shaped pottery drinking vessels, the Beaker people used copper implements, and they lived in a more communal fashion than their ancestors. According to the traditional theory, they migrated to Britain from the continent several thousand years after the land bridge disappeared, in numbers that overwhelmed the original settlers of the islands, but experts in the field now argue that the archaeological evidence does not support such a theory. It seems more likely that the Beaker people were part of the indigenous population who introduced new ways of life to an older, established community.

Regrettably, the Stonehenge we see today consists of only some of the original stones, combined with an unknown degree of reconstruction. Many of the stones have fallen, and a number have disappeared completely, some of them having been used in past centuries for building walls or for road repair. In 1993 and 1994, a careful study of the site was made using radiocarbon dating, which measures the amount of carbon-14, a radioactive isotope of carbon, present in organic material. Because carbon-14 has two more neutrons than ordinary carbon (carbon-12), it is unstable and subject to radioactive decay. The isotope is produced naturally in the atmosphere as a result of the Earth's bombardment by cosmic rays, and plants and animals absorb this radioactive carbon, which starts to decay once they die. Therefore, by measuring the amount of carbon-14 in a pick made of deer antler that was used to dig one of the ditches at Stonehenge, we can arrive at a reasonably precise estimate of when the ditch was dug. Of course, what we are really measuring is how long it has been since the deer died, but we can assume that the antler was recovered from the deer fairly promptly and then fashioned into a pick—and after five thousand years, a few years one way or another are all but meaningless.

Thus, even though the stones themselves cannot be dated, given that they do not breathe or otherwise absorb carbon dioxide, the tools that were used to shape them and put them in place sometimes can. In dating Stonehenge, researchers tested a variety of samples of bone and antler taken from different places around the site to see whether these would yield roughly consistent dates. What they discovered was highly revealing. Stonehenge was built in three separate stages stretching well over a thousand years, between approximately 3150 and 1600 B.C., and the site had a very different appearance at different phases of its construction. This discovery dealt a major blow to those who imagined that Stonehenge as we see it today was intentionally designed that way as a sort of giant neolithic computer. As it turns out, some of the features of Stonehenge that have generally been taken to be part of the computer were actually built at widely different times and quite possibly were never in use simultaneously.

That said, it must be stressed that the dating of Stonehenge necessarily entails a great deal of speculation, with the result that several rather disparate chronologies have been proposed. Even English Heritage refers to two completely different sets of dates in the material they provide about the site. The more traditional of these has the second stage of construction ending around 2100 B.C.,

What Our Forebears Knew

and the third stage beginning around 2000 B.C. But an alternative, and evidently more recent, chronology pushes these dates backward, with the second stage ending as early as 2500 B.C., and the third stage well underway by 2300 B.C. In addition, the sequence of building activity is unclear in several of its key stages, as are the specific time spans involved. Unfortunately, efforts at radiocarbon dating are not always successful. In some cases, no organic material is available for analysis. In others, the results may be inconsistent or discordant, either owing to contamination of the sample or because materials dating from different periods have intermingled. Or there may simply not be enough material to produce a definitive result. Not all samples yield meaningful conclusions when analyzed, and dates are regarded as reliable only when several samples from the same location agree among themselves. Where the chronology of Stonehenge is uncertain, I have generally adopted the dates provided by English Heritage.

Stage 1

The first stage of construction at Stonehenge, from about 3150 to perhaps 3000 B.C., consisted of the excavation of a large circular ditch, about 100 meters in diameter, inside which was a bank of earth. The circle of ditch and bank was broken by a large gap that faced in the direction of the midsummer sunrise. Over the centuries the ditch has gradually filled in, and the bank—which was certainly much taller when the ditch was first dug (although exactly how tall we cannot know)—has leveled to the point that it is now just 30 centimeters high. Deer antlers—which appear to date to 3000 B.C., give or take some sixty years—were discovered in large numbers inside the ditch and had evidently been used as picks for excavation and for breaking up the chalk. Rather than having been shed naturally, more than half the antlers were recovered from deer recently slain, suggesting that a sizeable quantity of antlers may have been required quickly for the construction project.

Inside the bank of earth, regularly spaced out around a circle about 80 meters in diameter, are the so-called Aubrey holes, named after John Aubrey, which appear to belong to the same period as the earthwork. First discovered around 1666, they are but shallow depressions today, although when they were originally dug they must have been fairly deep, for at one point they held wooden posts that formed a circle inside the ditch and bank. Over time, however, the Aubrey holes were gradually filled in with cremated bones and other remains. Unfortunately, nothing has been found among the cremated remains and other organic material that has proved usable for analysis, and so we have been unable to determine precisely when the holes were first dug, although they appear to belong to the same period as the earthwork. While it is generally assumed that the wooden posts were placed in the circle of Aubrey holes during this phase of construction, nothing survives of these posts today. We are able to infer their existence only from the observed structure of the holes themselves.

While it is not inconceivable that the Aubrey holes aided in tracking the Moon's movements (as we will see below), their purpose basically remains a mystery. Similarly, we cannot know why cremated human remains were placed in the holes. The existence, at nearby Woodhenge, of a grave containing the body of a small child has been interpreted to suggest that some kind of ritual burial, or perhaps even sacrifice, might have been involved. But the child's body had not been cremated, and in any case the link between the grave and Stonehenge is, at best, extremely tenuous. The idea that Stonehenge was routinely used for the ritual sacrifice of human beings or animals remains fundamentally a matter of speculation, unsupported by anything other than highly circumstantial evidence.

Stage 2

The second phase of building at Stonehenge, which stretched from approximately 3000 to 2500 B.C., initially involved very little construction as such. In fact, for many centuries this phase amounted more to an interlude in construction, characterized by some natural decay of the original earthworks, including the gradual filling in of the ditch (possibly the result of erosion) and by the partial filling in of the Aubrey holes. It appears, however, that toward the end of this period, around 2500 B.C., the first stones were brought to the site. These were the bluestones, which originally formed a ring inside the later Sarsen Circle and are today completely dwarfed by it. The stones weigh 4 to 5 tonnes apiece and were apparently brought from the Preseli Hills, in Pembrokeshire, situated in southwestern Wales—an extraordinary journey of some 380 kilometers. Part of this journey must have been carried out on rafts, across the Bristol Channel and then up the River Avon, past the site of modern-day Bristol, but much of it was necessarily overland through what was at the time heavily wooded country. Why such gigantic stones were brought from such a distant and inaccessible location is yet another mystery, but the effort involved was obviously immense. Perhaps stranger still is the fact that, after this period of intense work, Stonehenge was evidently abandoned for perhaps two centuries, leaving some of the bluestones yet to be put in place.

Stage 3

Most of what we see today belongs to the third and final stage of construction. Over a period of roughly four hundred years, possibly beginning as early as 2400 B.C., new sets of stones were added to the site. Because many of these cannot be reliably dated, the order in which they were set up is uncertain, and different sources give very different sequences of events. What is certain, however, is that this phase of building was marked by some truly colossal feats of engineering. Not the least of these was to dig up the bluestones and then place them in a new configuration—although this task pales in comparison to other, more spectacular additions to the site. Possibly the first set of stones to be erected, perhaps as early as 2300 B.C., was the outer ring of sarsens (a type of sandstone), now known as the Sarsen Circle. Of the thirty original stones, each roughly 4 meters tall, only seventeen are still standing. The circle of sarsens, 33 meters in diameter, was connected along the top by a lintel formed of stones that average 3.2 meters in length, although very few of these—only six—still survive.

The huge stones were brought to Stonehenge from a site on the Marlborough Downs, 30 kilometers to the north. Hauling them that distance was obviously an enormous undertaking, one that entailed the ingenious use of ropes and rollers, and even with that it is estimated that moving a single stone across a hill along the route would have required as many as six hundred men. We know that stone balls were used to hammer the sarsens into shape, and it is still possible to detect the marks in the ground where the stones were dragged across the site. As we can also see, the lintel stones were secured on top of the uprights by a system of mortises and tenons, or ball and socket joints, and were attached to one another by a similar arrangement consisting of a projecting tongue that fitted into a groove in the stone beside it. But we have no idea how these mammoth stones were raised into an upright position, much less how the equally massive lintel stones were lifted onto the top of them.

Inside the Sarsen Circle are the five enormous trilithons. Each is an arch made of two large vertical stones with a third laid across the top as a lintel, again held in place by ball and socket joints carved out of the stone. As usual, the radiocarbon dating is problematic, and all we can say is that these stones appear to have been set up somewhere between 2400 and 2100 B.C. and were probably erected after the Sarsen Circle was constructed. As in the case of the Sarsen Circle, though, their function is a mystery. Yet another stone, the so-called Altar Stone, which stands in the center of the circle, also came from southwestern Wales. Fashioned out of green sandstone, the Altar Stone measures 4.9 by 1.0 by 0.5 meters and weighs around 8 tonnes. In addition, there are the Station Stones, of unknown purpose, which are located just inside the earth embankment, roughly in line with the Aubrey holes, each of them surrounded by a ditch some 10 to 12 meters in diameter.

Later still the Y and Z holes were dug and the Avenue built. The latter is an earthwork some 2.8 kilometers long that extends in an arc from the northeast to the southeast of the central part of the site. It was probably constructed over a period of several centuries and appears to have functioned as a ceremonial approach to the site. The earthwork is not obvious to the casual visitor, however, nor is the area open to the public. The Y and Z holes form two additional circles at a distance of 11 and 3.7 meters, respectively, beyond the Sarsen Circle. The thirty Y holes, which were dug around 1600 B.C., are the youngest part of the site. The twenty-nine Z holes are of less certain age but seem to have been dug around 1900 B.C. (give or take 150 years or so) and are thus, like the Y holes, more recent than the stone structures. Although the two sets of holes were evidently intended as the emplacements for two additional stone circles, the stones were never put in place. Does this mean that Stonehenge was abandoned for a second time? Or did the use to which the site was put change in some unknowable way, so that the stones were never required? Or was some problem encountered that stopped the builders from finishing the site as they had intended? We do not know.

Was Stonehenge an Astronomical Computer?

The most ancient astronomical records to have come down to us are from China. For the most part, such records were carved onto oracle bones, and in some cases the original records were later transcribed (not always accurately) into volumes such as the *Shu-ching*. This text preserves the record of an eclipse of the Sun

What Our Forebears Knew

that may date back either to 1948 B.C. or to 2165 B.C., although in the opinion of some experts the eclipse was only a myth.[5] The first Chinese observation that can be reliably dated was an eclipse seen at An-Yang on June 14, 1330 B.C. At that time, work on Stonehenge had already been completed for about three hundred years, and the famous stones had been in place for some six to eight hundred years. In other words, the astronomical features of Stonehenge appear to antedate the Chinese records.

It is absolutely beyond doubt that Stonehenge had some sort of astronomical function. The famous alignment of the Heel Stone with the midsummer sunrise is proof positive of that. Theories have also been advanced regarding the alignment of certain of the stones with other astronomical events, particularly moonrise. Although these theories are less broadly accepted, they are not altogether implausible. Dozens of other possible alignments of the stones with the position of especially bright stars have been proposed, but these hypotheses remain controversial at best.

In addition to theories regarding the positioning of the stones, there is evidence to suggest that Stonehenge could have been used as a sort of calendar. The key to this is the Aubrey holes. There are fifty-six of them. This means that if a marker, such as a wooden post, were moved two holes per day, it would complete a full circle in 28 days—close to the 27.3 days it takes the Moon to revolve around the Earth. Intriguing though this is, however, the builders of Stonehenge could not possibly have known the length of the orbital period of the Moon around the Earth, for which they would have required a heliocentric model of the solar system and Newton's theory of gravity. Moreover, even if they had been aware of the period, the difference of 0.7 days (16.8 hours) between the 28 days needed to complete a circle of the Aubrey holes and the 27.3-day rotation period of the Moon would cause the calculations to lose accuracy rapidly. In just one year they would be off by a full 9 days and thus totally useless.

In fact, Neolithic people would have been considerably more interested in the cycle of 29.5 days from new moon to new moon or full moon to full moon. But to calculate the phases of the Moon with reasonable accuracy would require fifty-nine Aubrey holes, not fifty-six. Apart from the Moon's monthly phases, however, another period of time, known as a saros cycle, could possibly have drawn the interest of those responsible for the Aubrey holes. Saros cycles govern the periodicity of eclipses—that is, a lunar eclipse will repeat every saros cycle—with each cycle lasting 18.03 years. Once you are aware of the cycle, you need not perform complex mathematical operations to know that if an eclipse of the Moon occurred on such-and-such a date, there will be another exactly one saros cycle later. Furthermore, because solar eclipses also depend on the position of the Moon, saros cycles govern eclipses of the Sun as well as the Moon.

It is well known, of course, that solar eclipses have always inspired fear. In ancient China it was thought that a dragon was devouring the Sun, and one of the main jobs of court astronomers was to warn of impending eclipses so that the populace could be mobilized to come out into the streets and make loud noises, thereby scaring the dragon away. For early peoples, though, lunar eclipses were possibly even more frightening. Much depended on the reliable recurrence of the light of the full moon, which, among other things, ensured that any approaching predators or enemies could be seen. Knowledge of the Moon's periodicity also allows the seasons to be predicted and would thus have been critically important for agriculture. Imagine the feelings of Neolithic men and women when this dependable light in the sky was taken away and, as is often the case during a total lunar eclipse, the Moon turned blood red.[6] It would have been a terrifying spectacle to people who believed deeply in the powerful forces of nature but did not yet understand how these forces operated.

Had it been possible to use the Aubrey holes to keep track of saros cycles, then eclipses could have been predicted with a fair degree of precision. Might this, then, have been the purpose of the Aubrey holes—to mark the start and finish of a saros cycle? If the wooden marker were moved three Aubrey holes per year, it would complete a full circle in 18.67 years—but this is some eight months longer than one saros cycle. In fact, in order for this system to work for determining the beginning and end of a saros cycle, one would want fifty-four Aubrey holes instead of fifty-six. Moreover, knowledge of the saros cycles was not that widespread. The Chinese were familiar with them and in fact made extensive use of them for predicting eclipses. The Babylonians, however, whose culture was much more advanced than that of Neolithic Britain, were probably ignorant of saros cycles, except perhaps toward the end of their history. Could those who built Stonehenge, despite lacking the benefits of writing and sophisticated mathematics, have done better in this regard than the Babylonians? It seems pretty doubtful, and the existence of the two extra Aubrey holes makes it even less likely that this explanation is correct.

Explaining Stonehenge

Fascinating although it undeniably is, the evidence for Stonehenge as a tool for Neolithic astronomers is suggestive rather than conclusive. We can understand *why* the constructors of Stonehenge might have wished to predict eclipses, but it stretches credibility to say that this is the reason Stonehenge was constructed. In particular, the most obvious of its suggested astronomical uses—to calculate the summer solstice and thus fix the year, to monitor the progress of the seasons, or, just possibly, to determine when eclipses would occur—would have required the use of only relatively minor elements of the complex, notably the Altar Stone, the Heel Stone, and the Aubrey holes. If other, more striking features of Stonehenge, such as the sarsen stones, the trilithons, and the bluestones, had an underlying astronomical purpose, we have yet to figure out what it was—quite possibly because no such purpose ever existed.

To view the issue of purpose in its proper perspective, consider the span of time over which Stonehenge was constructed. Imagine starting to build a new football stadium in the year 2004, with the second phase of construction to take place in 2600 and the project to be completed in the year 3500. Yes, it can be done, but is there any guarantee that fifteen hundred years later people will want to play anything still recognizable as football in the stadium? In other words, the final use to which the stadium is put may have nothing to do with the original aim of its construction. Similarly, in its later stages of construction, Stonehenge may have been adapted to uses completely unlike those for which it was initially conceived.

In short, Stonehenge must have had multiple purposes, one of which appears to have been astronomical. But it was by no means the only one. It in fact seems more likely that Stonehenge served first and foremost as a ceremonial site, with its astronomical functions perhaps playing a secondary role. It could be that sacrifices were made to the gods during eclipses of the Moon, much in the same way that the Chinese made noises in the street. The ultimate aim would be to end the eclipse, whether by scaring away the Moon-devouring dragon or by placating the gods with sacrificial victims. (As an old colleague at the Bristol Astronomical Society, Bob Gilbert, used to put it, crudely but quite probably accurately: "They wanted to be able to predict eclipses so that they would know when to rip your heart out.") Similarly, the cult of Sun worship may have required that offerings be made to ensure the fertility of the land as it was warmed by the Sun at each summer solstice. Such might explain the cremated human remains found in the Aubrey holes.

Whatever it was, though, Stonehenge was clearly of immense importance to the Neolithic peoples of Britain. Extraordinary efforts were put into building it, and construction continued for well over a millennium. Out of what was at the time a very sparse population, hundreds, perhaps thousands, of people participated in its construction, expending vast amounts of energy to transport huge and heavy blocks of stone from distant quarries. One does not undertake such a massive, sustained effort without some overwhelming reason.

In fact, one of the obstacles to a clearer understanding of Stonehenge has been our rather arrogant tendency to assume that anyone who lived in the British Isles four thousand years ago is bound to have been an ignorant savage, too primitive and simpleminded to have had any deep purpose for creating such a structure. Granted, in many ways the Neolithic peoples of Britain were primitive. They had no tools beyond what could be fashioned from bone and antler, no written language, no large-scale agriculture, and no cities, whereas the Chinese and the Sumerians (the ancestors of the Babylonians) had all of these. But monumental sites such as Stonehenge—in their scale and their neatly dressed stone, as well as in the quality of engineering they demanded—were not the work of ignorant, unsophisticated people. It is nothing short of incredible that they were able to do what they did with no more than bone tools, picks made of antler, and an assembly of ropes and logs. Just imagine the efforts necessary simply to quarry and shape the 45-tonne sarsen stones using only bone, antler, and balls of rock, let alone transport them long distances and then raise them into a vertical position.

Certainly, then, Stonehenge was a sacred site and, at least in its final form, probably a temple of some kind. Some of its astronomical functions, such as its *possible* use as a tool for predicting eclipses, may have been closely associated with religious practices (although these functions would have involved parts of the site that were among the first to be abandoned). Others may have been directed toward more practical ends such as determining the date of the summer solstice in order to fix the calendar. Over the almost fifteen hundred years of its evolution, parts of the site would have been put to new uses, while others, as we have seen, were quite obviously neglected, perhaps permanently. In short, to think in terms of a single explanation of Stonehenge is to wish for the Moon. Likewise, much as we may crave answers to our many questions about Stonehenge, it is foolish to imagine that we will ever have them. No doubt, though, we will continue to ask them, and to be drawn into the mystery of the circles of stone that rise so abruptly above the Salisbury Plain.

SUGGESTIONS FOR FURTHER READING

Popular Articles and Books

Simon Welfare and John Fairley, "Circles and Standing Stones," in *Arthur C. Clarke's Mysterious World* (London: Book Club Associates, 1980).

> *Based on a series produced by Yorkshire Television, this article provides a serious and thoughtful introduction to some of the myths and realities of stone circles and megalithic monuments in general, as well as of Stonehenge itself. The article includes a short commentary by Arthur C. Clarke on the building of such monuments, which points out just how difficult it is for us to understand the motives of the constructors. For anyone seeking to understand the world of megaliths and stone circles, this article offers an excellent place to start.*

Francis Hitching, *Earth Magic* (London: Cassell & Co., 1976).

> *Again written in association with a television documentary, this book treads a fine line between archaeological and astronomical fact, on the one hand, and pseudoscience, on the other, in its effort to explore the findings of hard science in combination with more controversial issues. Solid information concerning stone circles and related phenomena such as ley lines—lines that can be drawn straight through a number of ancient, often sacred sites that for some reason are quite precisely aligned—is thus riddled with talk of mystical powers, connections with UFOs, numerology, and other such oddities. All the same, provided one reads chapters on subjects like dowsing with a skeptical mind (they are fascinating, but not to be taken very seriously), the book provides a great deal of useful information and background material.*

More Advanced Reading

Sir Norman Lockyer, *Stonehenge and Other British Monuments Astronomically Considered* (London: Macmillan, 1906).

> *Although this book is now very difficult to obtain, it is an important work by an acknowledged master of many historical aspects of astronomy, who in this case turns his attention to Stonehenge and its possible astronomical associations. Lockyer undertook a detailed analysis of the alignment of certain stones with the midsummer solstice and concluded, as did other contemporary astroarchaeologists, that Stonehenge was a temple to the Sun. Although in light of subsequent research this conclusion seems less plausible today, Lockyer's work continues to be of great value.*

R. M. J. Cleal, K. E. Walker, and R. Montague, *Stonehenge in Its Landscape: The Twentieth-Century Excavations*, English Heritage Archaeological Reports no. 10 (London: English Heritage, 1995).

> *A heavyweight tome in all senses, this book is the definitive scientific study of Stonehenge. It discusses all the twentieth-century archaeological studies of Stonehenge, including those using radiocarbon dating. In no sense is it an easy read, but it contains a wealth of detail for people who wish to delve more deeply into the scientific studies that have been made of the site.*

On the Internet

NOTE: One of the blessings of the Internet is that information about almost anything can be found somewhere if you know where to look—and one of the curses is that because anyone can post information about any topic they wish, it is hard to be sure whether the material you've located is accurate. Heaped around the gems is a great deal of rubbish. Another problem, of course, is that Web addresses change constantly. What follows here, and likewise in the subsequent chapters, are a few carefully selected Web sites that not only offer valuable information but can also be used as a starting point for further searches. The addresses provided here were current as of March 2005. Remember that such addresses are often case-sensitive, so be sure to type the address exactly as it is given.

Earth Mysteries: Stonehenge
http://witcombe.sbc.edu/earthmysteries/EMStonehenge.html

> *Despite the fame of Stonehenge, there are surprisingly few good Web sites where one can find reliable information and high-quality pictures. This page—the work of a lecturer, Chris Witcombe, at Sweet Briar College, in Virginia—is not only beautifully mounted but well researched and documented, and it contains a number of carefully executed maps, diagrams, and photographs that are superbly informative. The site covers the history, construction, and use of Stonehenge, offering an abundance of information presented in an orderly and very accessible way. Although one needs to remember that, as is so often the case with sources on Stonehenge, part of the content is informed speculation and should be treated as such, the Earth Mysteries site is an excellent place to learn about Stonehenge.*

Stonehenge. Official English Heritage Web site
www.english-heritage.org.uk/stonehenge/

> *Until fairly recently, the official English Heritage Stonehenge site was so poorly executed as to be almost useless—a few not-very-good photographs, a scientific study on the dating of Stonehenge that threatened to be incomprehensible to most readers, and no background or general information of any kind (not*

even the site's hours or the price of admission). Happily, a completely new Web page has been prepared that tackles all these issues. It offers reasonably extensive information on Stonehenge and its history (although still less than can be found on some of the other, unofficial, sites), as well as on visits and access, including detailed instructions about how to get to Stonehenge. There is also a new photo gallery containing some fifteen small but quite high-quality images.

Stonehenge
www.britannia.com/history/h7.html

A single page, but well presented, that provides simple, straightforward, and accurate information. It is a pity that the site includes so few photographs, but as a brief introduction to Stonehenge it is definitely worthwhile. Along with the two mentioned above, one of the few Stonehenge sites worth a look.

CHAPTER 2

How Did the Stars Get Their Names?

When we look up at the night sky on a clear, moonless night, we see the darkness speckled with what seems like millions of stars. But even a person with very good eyesight, living far from city lights, will see only about two thousand stars on any particular night—although this alone is enough to produce a bad case of vertigo. People often look at the enormous multitude of stars and wonder whether anyone has ever counted them all and whether they all have names. Capitalizing on this ignorance, a number of companies have started to "sell" stars, banking on the suggestion that naming a star after a loved one makes an ideal present. These companies do not point out that most of the stars bright enough to be seen with the naked eye were named long ago. Even the very faintest stars, visible only with a large telescope, have at least catalogue designations.

The naming of the stars has a long history, dating back over twenty-five hundred years in the West. In fact, perhaps as much as four thousand years ago the Chinese had given names to the stars, and the Egyptians may have done so much earlier still, although no record of their work has come down to us. The constellations—groups of stars that appear to form figures in the sky—are generally known today by their Latin names but were for the most part named by the Greeks, although the constellations they recognized were often borrowed from the Babylonians. The Greeks also named many of the stars themselves, although most of them now have Arabic names, while the first modern catalogue of the stars, compiled by Johann Bayer and published in 1603, mixes both Latin and Greek. Then, some four hundred years ago, a Danish astronomer produced a map of the sky, based on his own observations, in which he included a completely new constellation. His creation filled in a region in the old sky maps that was sprinkled with rather faint stars but otherwise vacant. He named the newly formed grouping himself, following tradition by choosing a name based on classical Greek mythology, and thus became the first person to add a new constellation to the forty-eight listed over a millennium earlier by Ptolemy.

In so doing, he initiated what was quickly to become a major trend that left astronomers from many countries fighting for the chance to leave their mark on the sky, in much the same way that their countries competed to explore and colonize the world. This celestial competition ended up getting somewhat out of hand, resulting in a bewildering array of maps of the sky that listed many new constellations. Again, these new constellations generally consisted of clusters of fainter stars that were not part of one of the familiar groupings and thus served to fill up small, largely empty areas tucked in among the traditional constellations. But with each map recognizing a different set of groups, and in some cases different names for them, it was clear that someone needed to step in and bring order to the celestial chaos. In the end it was the International Astronomical Union (IAU), the "government" of astronomers, formed in 1919, that took up the matter. It was not until 1933 that the IAU began to regulate the naming of stars and constellations, after a commission set up to investigate the problem had reported its conclusions and these had been studied and ratified. Today, the naming of stars is extremely strictly controlled, and many of the unjustifiable additions to the sky made in the seventeenth and eighteenth centuries have been removed.[1]

Mapping the Night Sky

Many ancient civilizations named and probably also mapped the stars. Certainly the Chinese did, as did the Egyptians and the ancient Mesopotamian peoples—the ancestors of the Sumerians, the Babylonians, and the Assyrians. Because written records are extremely scarce, though, it is almost impossible to determine who deserves credit for the earliest astronomical observations and maps. Moreover, what few written records do exist can rarely be dated with any precision. The first record to which a reasonably reliable age can be assigned is the famous Venus tablet, which describes a set of observations of the planet Venus made by the Assyrians somewhere between 1700 and 1600 B.C., during the reign of King Amisaduqa. Given that the Babylonians (and also the Chinese) were in the habit of copying and recopying their records, the Venus tablet itself actually dates to the seventh century B.C., however, making it roughly a thousand years younger than the observations it describes. Precious few records survive from the period prior to the sacking of Babylon in 689 B.C., and it is increasingly unlikely that any more will be found. Although the city was rebuilt, and its culture again flourished, our knowledge of Babylonian astronomy remains scant. What little we do know, however, tells us that many of the constellations they identified are akin to our own. The Babylonians had a constellation called the Lion and another called the Bull, which correspond to the modern-day Leo and Taurus. In fact, the zodiac itself goes back to the Babylonians. Like the word *zoo,* the term *zodiac* derives from the Greek *zōion,* "living being," and the original Babylonian zodiac was indeed a circle of animals around the sky.

Like the Babylonians, the Egyptians also took considerable interest in the sky. The Great Pyramid at Giza is thought to have been built between about 2590 and 2565 B.C. for the pharaoh Cheops, or Khufu, the second king of the Fourth Dynasty. The pyramid, also known as the Pyramid of Cheops or the Pyramid of Khufu, is carefully aligned with what was the Pole Star at the time the pyramid was built. Because of the wobble of Earth's pole, which has a 26,000-year period, the Pole Star—that is, the star in the northern hemisphere toward which the axis of the Earth points—changes over time. At present it is Polaris, or the North Star, but forty-five hundred years ago it was a rather fainter star called Thuban, in the constellation of Draco (the Dragon). It is with this latter star that the Great Pyramid is aligned. Nothing was known about the Egyptian sky maps, however, until Napoleon's army found one in Egypt in 1798. Quite possibly the first Egyptian maps of the sky predate the Assyrian observations recorded on the Venus tablet, but we will probably never know for sure.

The third of the great historical cultures to make a serious study of astron-

What Our Forebears Knew

omy was the Chinese. Here we have more information because records of many ancient Chinese observations of the sky have survived and in fact form a veritable treasure trove for astronomers. In fact, up to about 1600 A.D., when Western records began to improve, almost all the best astronomical data were Chinese. The oldest surviving Chinese sources date back to around 2000 B.C., so there is good reason to believe that the Chinese had mapped the sky at least by then, and possibly much earlier. We know a substantial amount about the Chinese constellations and about the names they gave the stars because they made extensive use of the heavens in their horoscopes and in foretelling the future on the basis of celestial omens. The Chinese constellations differed from the Western ones, however. The Chinese divided the sky into almost three hundred small groupings of stars that we would today call asterisms. Some of the Chinese constellations have picturesque names—the Wagging Tongue, for example—but others, such as the Cow Herder or the Dogs, have a more familiar ring to them and are indeed similar in name, if not in form, to the constellations recognized internationally today. We have the Herdsman (Boötes) and the Great Dog and Little Dog (Canis Major and Canis Minor), as well as the Hunting Dogs (Canes Venatici), but the configurations of stars to which these names refer are completely different from the group of stars that the Chinese called the Dogs. Because the Chinese identified so many groups of stars—about four times as many as we do—they often subdivided what are to us the standard constellations. A good example of this is the W formed by the five brightest stars in Cassiopeia, which was split into no fewer than three parts in Chinese sky maps. The Chinese also had an equivalent of the zodiac, but it was divided into twenty-eight lunar mansions (the groupings of stars through which the Moon passes each month) rather than into the twelve constellations of the Western zodiac. Although the lunar mansions each overlapped with some part of one or more of our signs of the zodiac, Chinese astrology was fundamentally unlike our own.

But the Chinese, the Egyptians, and the Mesopotamian peoples were far from the only ones who took an interest in the night sky. The Aztecs, the Incas, and the Mayas all studied the stars, as to some extent did other, less well-known cultures. As we have seen, the builders of Stonehenge apparently aligned some of the stones with celestial landmarks, notably the Sun and the Moon. The second phase of construction at Stonehenge took place roughly between 3050 and 2500 B.C. and is thus far older than the earliest Babylonian observations. The so-called Aubrey holes, which may have been used for predicting lunar eclipses, belong to that phase.

No astronomical records of any kind exist, however, so as far as the early British Isles are concerned, we have to make do with archaeological findings coupled with a great deal of guesswork. In a large degree the same comments also apply to Native American stone formations such as the famous Big Horn Medicine Wheel in Wyoming. This rock work consists of a central cairn or rock pile surrounded by a circle of stones. Twenty-eight lines of cobblestones link the central cairn, which is 4 meters in diameter, with the surrounding circle, which is some 25 meters across, with the result that the whole structure looks rather like a wagon wheel. Many similar structures exist, most of them in the Canadian province of Alberta, some of them more than 12 meters in diameter. As the use of the term *medicine* suggests, these structures must have had a religious significance to their builders, possibly as a ceremonial site, although we do not know their original purpose (or purposes). It also appears that, as in the case of Stonehenge, the ritual or ceremonial use of these sites may have changed over time. Some, like the astronomer John Eddy, have argued that specific rocks or cairns were positioned so as to align with the midsummer sunrise or with certain especially bright stars, although these astronomical theories tend to be regarded with skepticism. All the same, the central cairn of the Big Horn Medicine Wheel and one of the peripheral cairns do seem to line up very accurately with the midsummer sunrise, in the same way that the outlying Heel Stone at Stonehenge does when viewed from the center of the circle.

The Mysterious Greek Sky Mapper

We do not actually know how the most ancient Greeks divided the sky into constellations. Homer made reference to several of the constellations and some of the more prominent asterisms in the *Iliad*, which was probably composed in the seventh century B.C., and the Greek poet Aratus of Soli provided a verse description of forty-four constellations in his *Phaenomena*, written during the third century B.C. But it was not until approximately 150 A.D. that the names of the Greek constellations were formally laid down in writing. The person responsible for the most extensive written record of the constellations is a somewhat mysterious figure. Although this may not have been his original name, we know him as Ptolemy or, to give him his Latin name, Claudius Ptolemaeus. Ptolemy was Greek, but, as one might infer from his name, he was born in Egypt, which boasted a relatively large Greek population at the time. He lived during the second century A.D. and appears to have resided in Alexandria, where he may well have worked in the famous Alexandrian library.

During the second century B.C., an astronomer known as Hipparchus of Nicaea had compiled a catalogue that listed the names of about 850 stars. His work no longer survives, but it was probably completed around 130 B.C. Otherwise, very little is known about his life, and all but one of his works have been lost. Ptolemy set out to revise and expand the catalogue of Hipparchus and to improve on Hipparchus's calculations of the movements of the planets, which he did with great success. He wrote a work called *Mathzēionmatikē Syntaxis*, or the *Mathematical Compilation*, generally known today as the *Almagest*. Ptolemy's work is an astronomical tour de force divided into thirteen books. Books 1–6 and 9–13 analyze the motion of the various bodies in the solar system and map out the structure of the universe, with the Earth at the center of the sphere of the stars. The Earth itself is surrounded by spheres within which the Sun, the Moon, and the planets rotate in small circles of their own, called epicycles. Now familiar to us as the Ptolemaic system, it was regarded as definitive for some fifteen hundred years after it was laid out. In books 7 and 8 Ptolemy described what he called the fixed stars, to distinguish them from the wandering stars, that is, the planets. Here Ptolemy furnished a list of about 1020 stars, noting the position and brightness of each one. The stars were assigned one of six magnitudes on the basis of how bright they seemed to Ptolemy's eye, the brightest being magnitude 1 and the faintest magnitude 6. With certain minor modifications, this classification is still in use today.[2]

Ptolemy divided the sky visible from North Africa into forty-eight constellations. Of these, twelve were in the zodiac and the other thirty-six were scattered over the rest of the sky. These constellations are often referred to as "Ptolemy's forty-eight originals" to distinguish them from later additions. A list of these constellations is provided in Table 2.1. We do not know how many of Ptolemy's constellations were genuinely Greek, in the sense that it was the Greeks who first put names to them. But we do know that the Babylonians had already defined the twelve signs of the zodiac, and it is likely that many of the other constellations in Ptolemy's sky were also borrowed from the Babylonians. The source of Ptolemy's list of constellations is reputed to have been a poem, long since lost, written by the mathematician and astronomer Eratosthenes some five hundred years before Ptolemy produced the *Almagest*. By one estimate, as many as thirty of the forty-eight "original" constellations are Babylonian in origin.

Table 2.1 Ptolemy's forty-eight original constellations

Northern hemisphere	Zodiacal	Southern hemisphere
Ursa Major	Aries	Cetus
Ursa Minor	Taurus	Orion
Draco	Gemini	Eridanus
Cepheus	Cancer	Lepus
Boötes	Leo	Canis Major
Corona Borealis	Virgo	Canis Minor
Hercules	Chelae Scorpionis	Argo Navis
Lyra	Scorpio	Hydra
Cygnus	Sagittarius	Crater
Cassiopeia	Capricornus	Corvus
Perseus	Aquarius	Centaurus
Auriga	Pisces	Lupus
Ophiuchus		Ara
Serpens		Corona Australis
Sagitta		Piscis Australis
Aquila		
Delphinus		
Equuleus		
Pegasus		
Andromeda		
Triangulum		

There is an old rhyme that helps one remember the names of the signs of the zodiac:

The Ram, the Bull, the Heavenly Twins,
and next the Crab, the Lion shines,
the Virgin and the Scales,
the Scorpion, Archer, and the Goat,
the Man who holds the Watering Pot
and Fish with glittering tails.

What Our Forebears Knew

In the modern-day zodiac, Libra, the Scales, is anomalous in being the sole inanimate object. In the zodiac of Ptolemy, however, Libra has disappeared and in its place is an interloper called Chelae Scorpionis, or the Scorpion's Claws. In other words, in the Ptolemaic zodiac Libra was an animal like all the others, or at least part of an animal. Why and by whom the change was made is not clear, but it dates back many years.

Two other constellations have changed as well. Argo Navis, or the Ship Argo, was originally by far the largest constellation in the entire sky. By the nineteenth century, however, Argo Navis had been split into three separate constellations: Carina (the Keel), Puppis (the Poop Deck), and Vela (the Sails). Somewhat confusingly, though, the original name and the three new ones were used in tandem until the IAU finally made the division into three parts official in 1933. The other change is one of detail. Although it is still listed as a single constellation, Serpens (the Serpent) is now divided into two parts, Serpens Caput (the Serpent's Head) and Serpens Cauda (the Serpent's Tail), which are separated in the sky by the constellation of Ophiuchus. The three have sometimes been taken as a single grouping, Serpentarius, that shows a man wrestling with a serpent, which has wrapped itself around him. As the IAU sees it, though, Ophiuchus has won the fight, and the serpent has been split in two, one half lying on either side of him.

Heros and Villains in the Skies

All of Ptolemy's constellations are associated with Greek myths, although some, perhaps many, of the legends were assimilated from the Babylonians. It was Eratosthenes, however, who established the links between the constellations and the various Greek legends in his *Catasterismi*, composed in the third century B.C. For the Greeks, many of the constellations recalled the deeds of great heroes, while other constellations were thought to have been created as a reward from the gods: the honor of a place in the heavens. Or, in the case of those who trifled with or otherwise offended the gods (always a bad idea), being immobilized in the firmament could be a punishment—or at least a dubious favor, as the story of Gemini, the twins, illustrates. Castor and Pollux were the twin sons of the king and queen of Sparta. Pollux was immortal, but Castor was not, and so when Castor died, Pollux begged Jupiter to allow him to share his immortality with his dead brother. No doubt to his chagrin, although Jupiter agreed to bring Castor back to life, it was at the cost of placing both youths in the heavens, where their immortality was of very limited benefit. One can well imagine fathers looking up at the stars and telling their children this and other stories that the constellations brought to mind, either to inspire them to venture to great deeds themselves or to warn them of how great the wrath of their gods could be.

Orion, the Hunter

Three great heroes appear in Ptolemy's sky, Orion, Hercules, and Perseus, although the three have very different histories. In the mythology of the Greeks Orion was first and foremost a mighty hunter. Two versions of his story have come down to us. In one, Orion was the son of Neptune and the queen of the Amazons. Inheriting his mother's prowess as a hunter, he boasted that he could get the better of any animal on Earth. Apollo, worried by his sister Diana's infatuation with Orion and angered by his boasts, sent a scorpion to sting his heel and kill him, thus demonstrating that Orion's claim was an empty one. On his death Orion was placed in the sky, and Scorpio, the animal he could not defeat, was placed on the opposite side of the sky, where it could harm him no longer.

An alternative version of the story also has Apollo as Orion's nemesis. Orion lived with Diana, and they shared a life of hunting. Their relationship was such that Apollo feared his sister would lose her virginity and resolved to end the liaison by any means. One day, walking along the seashore with Diana, Apollo spied Orion, who had waded so far out to sea that only his head was visible above the water. Apollo challenged Diana to hit this distant target, which his sister took to be a rock. Diana's aim was true, and her arrow killed Orion. Devastated by grief, Diana placed her platonic love in the sky, where he would be safe in the future.

The magnificent constellation of Orion is visible in the winter sky and in the spring, when it sets in the west in the early evening. We find Orion accompanied by his hunting dogs, Canis Major and Canis Minor (the Great Dog and the Little Dog). Under his feet is a hare (Lepus), to mark his fondness for hunting rabbits, and in one hand he holds his club aloft, ready to strike the bull (Taurus) that is charging him.

Hercules

Perhaps the greatest of all heroes, Hercules was the son of Jupiter and Alcmena, in one of Jupiter's many extramarital excursions. Hercules had a twin brother, Iphicles, who was a mortal, although Hercules himself was half immortal. Juno, Jupiter's wife and queen of the gods of Olympus, was extremely displeased with her husband's affair and set out to take revenge on Hercules himself. She sent two snakes to kill him in his crib, but, as if to lend a taste of what was to come, the infant Hercules strangled them with his bare hands—one in each hand. Tormented constantly in his adult life by Juno's persecution, Hercules was finally driven to madness and in a fit killed his wife, Megara, daughter of Creon, king of Thebes, and their children. When he realized what he had done, he was horrified. He consulted the oracle at Delphi and was told that to atone for his crime he must serve his uncle and lifelong rival, King Eurystheus, for twelve years. It was Eurystheus who set him the twelve monumental labors that would finally earn him his pardon.

His first task was to kill the lion of Nemea, which he wrestled to death in its den, skinning it and then donning its impenetrable hide to wear from then on as a protective cloak. Next he was ordered to kill the Lernean Hydra, a giant poisonous sea snake adorned with many heads. Every time that Hercules cut off one of its heads, it would grow two new ones, becoming ever more powerful and dangerous. But the clever Hercules finally burnt the stumps before they could grow back. During this battle Juno sent a crab to aid the Hydra, which clawed at Hercules persistently until he killed it by crushing it under his foot.

Several of his tasks were gruesome. The Stymphalian birds, which he killed, the mares of Diomedes, which he captured, and cattle of Geryon, which he stole, all ate human flesh. To capture the Erymanthian boar he was obliged to kill Pholus and Chiron, two centaurs. Other tasks were also suitably Herculean, such as the year he spent chasing the Cerynean hind before finally capturing it, or his cleaning of the Augean stables, which were filled with thirty years' worth of muck from three thousand oxen. His mission was to clean them in a single night, a task he successfully carried out by the simple expedient of redirecting the rivers Alpheus and Peneus so that they flowed through the stables. He was also ordered to capture the Cretan bull, but that feat seems rather prosaic in comparison to the more unusual perils he faced in stealing the girdle of the queen of the Amazons, Hippolyta, and in venturing into the underworld to abduct Pluto's dog, Cerberus, which pitted him against a powerful god. His final task was to seize the golden apples from the garden of the Hesperides by first killing the dragon that guarded them.

Hercules also accompanied Jason on his quest for the golden fleece. He eventually married a second wife, Deineira, only to die after she accidentally poisoned him—although in some versions of the myth he died after putting on a poisoned cloak sent by Juno. In the end Hercules was immortalized by Jupiter, who placed him in the sky, possibly to demonstrate to Juno the error of her ways. Many of the players in the story of Hercules' labors also have places in the heavens. The lion and the crab are, of course, Leo and Cancer, and the multiheaded Hydra and the dragon Draco are also constellations. (In fact, Hydra is the largest constellation in the sky.) Both Centaurus and Sagittarius are centaurs, perhaps representing Pholus and Chiron, the two centaurs Hercules killed while capturing the Erymanthian boar. We also find Jason's ship, in the form of the constellation Argo Navis, as well as the object of his quest, the golden fleece, which appears as Aries, the ram.

Perseus

Perseus is in some ways the original all-Greek hero, the clean-living good guy, doer of virtuous deeds. The son of Jupiter and Danae, a mortal woman, Perseus had been charged with killing the Gorgon Medusa, whose hair consisted of a tangled mass of writhing serpents. Unfortunately, Medusa had a particularly effective defense against attack: if you looked at her face you were immediately turned into stone. The supply of young heroes willing to fight her had thus grown sadly depleted, as many of them now decorated the entrance to her cave. Perseus, however, as the son of Jupiter, counted on willing assistance from the gods of Olympus. He received a pair of winged sandals from Mercury to allow him to fly, a helmet from Pluto that would render him invisible so that he could approach Medusa and her two Gorgon companions undetected, and a highly polished shield from Minerva. Perseus was also warned never to look directly at Medusa but instead to be guided by her reflection in his shield while creeping up to slay her. This he duly did, and he cut off her head as a trophy, which he kept in a bag at his waist.

While Perseus was out on his quest, a tragedy was developing in the kingdom of Ethiopia. Queen Cassiopeia was intensely proud of her daughter, Andromeda, and boasted that her beauty was greater than that of the sea nymphs, the Nereids—the daughters of Neptune, the god of the sea and second only to

Jupiter in his power. Neptune was, like most of the gods of Olympus, both bad tempered and vengeful. Hearing of this insult to his daughters, he decided to send a sea monster to ravage the land. Cassiopeia's husband, King Cepheus, witnessed the destruction of his country with increasing despair and finally resolved to consult the oracle at Delphi in hopes of discovering a remedy for the situation. The pronouncements of the oracle were, unfortunately, rarely to the liking of those who solicited them. True to form, the oracle informed King Cepheus that it was his fate to chain his daughter to the rocks on the seashore and leave her as a sacrifice to the monster. This he very reluctantly did.

Just as the monster was approaching to claim its prize, Perseus was flying overhead on Pegasus, his winged horse. He saw Princess Andromeda's tragic plight and, in the best tradition of heroes, arrived at her side just in the nick of time. Taking the Gorgon's head out of his bag, he turned the monster to stone before setting Andromeda free. As a reward, the grateful King Cepheus and Queen Cassiopeia offered Perseus Andromeda's hand in marriage. At his death, Perseus was also rewarded by the gods with a place among the stars. In fact, all the players in this story now reside in the sky. On a late winter evening, if we look to the west and northwest, we can see Perseus alongside his bride, Andromeda. Above both of them is Queen Cassiopeia on her throne, with the hand of Perseus stretched out toward it. (According to Roman legend, though, Cassiopeia was placed in the sky on her throne not as a reward but as a punishment for her boastfulness. In the course of her circuit around the heavens she is often left hanging rather uncomfortably, upside down.) Between Cassiopeia and the Pole Star is King Cepheus. Separated from Andromeda by a stretch of heavenly waters that carries Pisces (the Fishes), we find the Cetus (the Whale), originally the sea monster but who has, over the years, evolved into a somewhat less threatening cetacean.

Latin or Greek?

In recounting these myths today, we almost always use the Roman equivalents of the Greek names: Jupiter rather than Zeus, Neptune rather than Poseidon, Hercules rather than Heracles, Juno instead of Hera, and so on. But the stories are the same in both the Greek and Roman traditions. As is well known, the Romans borrowed extensively from Greek culture, and thus the Roman gods and myths were closely similar to the Greek. Although the Roman Empire never formally annexed

Greece, the Romans conquered the Greeks in the Third Macedonian War, which ended in 168 B.C., and subsequently exercised firm control over Macedon, now divided into four republics. As the Roman Empire spread across the Western world, Latin naturally became the dominant tongue. Even after the fall of Rome, Latin remained the language of culture, of philosophy, and of scholarship. With the rebirth of scientific inquiry around the start of the fifteenth century, Latin also became the language of science, including astronomy.[3] The constellations thus came to be known by their Latin names, which we still use today. In astronomy, as in most sciences, tradition can be a powerful agent.

Are There Fifteen Signs of the Zodiac?

A number of years ago, the Royal Astronomical Society (RAS) created a great stir by announcing the existence of a thirteenth constellation of the zodiac, Ophiuchus. This was hailed in the media as a major new astronomical "discovery" and became a moderately large news story, with CNN and various other international news media reporting on it in some detail. But it is not news that Ophiuchus intrudes into the zodiac, cutting in between Scorpio and Sagittarius, and that the Sun, Moon, and planets can appear to be "in Ophiuchus—in other words, that they can be situated within the boundaries of the constellation." The RAS's announcement was one of the periodic skirmishes between astronomers and astrologers. How effective these are in convincing people that astrology is a pseudoscience is matter of debate. Witness the rather frightening statistic that in the United States, arguably the most advanced country in the world, there are ten times as many professional astrologers as astronomers.

What may be more surprising to many readers is that the RAS's announcement perhaps did not go far enough. Although only thirteen constellations are crossed by the ecliptic—the line in the sky that traces the path of the Earth's orbit around the Sun—the Moon and the planets are not limited in their movement precisely to this line. Rather, because their orbits are tilted slightly with respect to the Earth's, the Moon and the planets can climb a few degrees north or south of the ecliptic.[4] Thus the Moon, as well as some of the planets, can also be "in" two additional constellations that do not belong to the zodiac. If you doubt this, look up the position of Saturn between March 29 and 31, 1999: it was in Cetus. This curiosity of celestial geography does not seem to have been picked up by

What Our Forebears Knew

astronomy books, although many of them refer to Ophiuchus's "invasion" of the zodiac. Cetus is close enough to the ecliptic, however, that the Moon spends a respectable amount of time there. When the Moon runs south of the ecliptic in this region, it enters Cetus twice each month, at the points where this constellation wraps around Pisces. In late March 1999 Saturn also clipped the northern border of Cetus, to the east of Pisces. Since then, though, the planet has been far enough north to avoid other unexpected encounters of this kind.

The other zodiacal intruder is Orion. The upraised arm of the hunter extends upward to a point just south of the ecliptic to the south of Gemini. This narrow northward extension is sufficient to allow the Moon and the planets to enter Orion, if only briefly. On New Year's Day 1999, the Moon was actually in Orion for about ten hours, from 05:00 to 15:30 UT. (Universal Time, or UT, is the designation now generally used among scientists in place of Greenwich Mean Time, or GMT.) Beginning then, the Moon crossed this constellation each month until the progressive northward shift of its monthly path, caused by the saros cycle, moved it away from Orion in the year 2002. Nor is the Moon alone in entering Orion. From June 4 to 12 and again from July 10 to 14, 2001, Mercury did so, and the following year Saturn spent almost all of September and October in Orion. This must have provided an interesting conundrum for astrologers. A few days in a nonzodiacal constellation can probably be dismissed as an accident; two whole months takes some explaining.

So the next time that you read in your horoscope that the Moon, or Mercury, or Saturn is in Gemini or Pisces, check a star map. Perhaps you are being misled. It may be that the planet is really in Ophiuchus, or even in Orion, or Cetus. So the question remains. Are there only twelve constellations of the zodiac? Should there be thirteen? Or even fifteen?

How the Stars Got Arabic Names

As we have seen, although we use their Latin names today, the names given to the constellations are largely of Greek origin. For the greater part, however, star names are most definitely not Greek, nor are they Latin. Indeed, it's not hard to notice that names like Aldebaran, Alphard, Algol, Rasalhague, Zubenelgenubi, Deneb al Giedi, and Betelgeuse have an Arabic ring to them. In fact, apart from a few modern names and the names of a handful of familiar stars such as Castor and Pollux, the mythical twins placed by the gods in the heavens as the constellation of Gemini, almost all the proper names of the stars are Arabic. Some of these names conjure up an image of the star. For example, the name Alphard is really "al-Phard"—the Solitary One. Visually, the name is appropriate, given that Alphard is by far the brightest star in Hydra and is situated in a barren area of the sky close to the North Galactic Pole, where there are few stars, bright or otherwise. Alphard is easy to recognize because no other stars of comparable brightness lie nearby, and, fittingly, Alphard is the only star in the constellation of Hydra that was dignified with a proper name.

Other names describe the location of the star in question. Aldebaran means "the eye of the bull," which describes its position in the constellation where it is found: it is the star that marks the eye of Taurus, the bull that is charging Orion. Betelgeuse means "the shoulder of the giant," and, sure enough, it marks the right shoulder of the figure of Orion. Several stars include the word *deneb*, "tail," in their name. The name of the star in the tail of Cygnus, the Swan, is simply Deneb, while other stars have names that consist of "Deneb" plus some other term. Denebola, for example, means "the tail of the lion."

As we have seen, included in Ptolemy's *Mathematical Compilation* was a detailed list of the constellations and their stars. But Ptolemy wrote in Greek, not Arabic. The original Greek manuscripts were lost many centuries ago, many having perhaps perished when the library of Alexandria was destroyed. However, the *Mathematical Compilation*, which had come to be known as the *Great Compilation*, survived in an Arabic translation called *al-Majisti*, literally, "the Greatest." Ptolemy thus owes his legacy to the work of a largely forgotten translator. The names Ptolemy had given to the stars, such as "the eye of the bull" or "the shoulder of the giant," were accordingly rendered into Arabic, eventually giving us Aldebaran and Betelgeuse. *Al-Majisti* was subsequently translated into Latin as *Almagest*. After it was composed, Ptolemy's catalogue was the standard reference work for astronomers for nearly fifteen hundred years.

Letters and Numbers

Many, but by no means all, of the brightest stars in the sky have individual names such as Alphard or Aldebaran or Betelgeuse. Besides these traditional names, however, most astronomy books include other names that begin with a number or a Greek letter, followed by a Latin name. Thus, Betelgeuse is also listed as 58 Orionis and α Orionis, Aldebaran is 87 Tauri and α Tauri, Pollux is 78 Geminorum and β Geminorum, and so on. Where did these other names come from?

In 1603 a German astronomer named Johann Bayer published a sky map and catalogue called *Uranometria*. Faced with the problem that so many of the stars lacked individual names, Bayer turned to the constellations. Each star in a constellation was assigned a Greek letter, which was attached to the possessive form of the Latin name of the constellation. The brightest star in each constellation was designated alpha (α), the second brightest beta (β), then gamma (γ), delta (δ), and so forth. Thus Alpha Tauri is the brightest star of Taurus. For the most part, Bayer's system works reasonably well, but there are some curious exceptions. For example, the brightest star in Sagittarius is Epsilon Sagittarii, followed by Sigma Sagittarii—whereas Alpha Sagittarii is the fifteenth brightest. In Ursa Major, the relative brightness of the stars is ignored, and they are instead designated in strict order of their position, from the front end of the Dipper along its handle (Alpha Ursae Majoris) to the tip of the bear's tail (Eta Ursae Majoris). A few of Bayer's stars are not even stars, as in the case of Omega Centauri, which is actually a magnificent globular cluster of stars. But despite these quirks, the Bayer system is still very popular and is widely used in popular astronomy books and by astronomers everywhere.

An alternative system for cataloguing bright stars was developed by John Flamsteed, England's first Astronomer Royal, who was charged with finding a definitive way to fix a ship's position at sea by using the movement of the Moon among the stars. Flamsteed made a catalogue of the sky and, moving from east to west, allotted a number to each of the stars in a constellation. Thus, the westernmost star of Ursa Major, Muscida (which is Omicron Ursae Majoris in Bayer's system), was designated 1 Ursae Majoris by Flamsteed. Although the Flamsteed numbers are less commonly used than Bayer's Greek letters, they were a forerunner of the system employed by modern catalogues, which also use a star's position as the basis for its designation.

Some Stellar Curiosities

Not all the names of stars are Arabic. In fact, there are a number of curiosities among the names officially accepted today. Take, for instance, the star in Canes

Venatici (the Hunting Dogs) called "Cor Caroli," a Latin name meaning "Charles's Heart." It was added to the sky in 1725 by the astronomer Edmond Halley, a staunch royalist, to honor his patron, Charles II. Nor did Halley's efforts stop there. Around 1680 he proposed adding a constellation that he christened Robur Caroli, "Charles's Oak," to commemorate the oak tree that Charles supposedly hid in to escape his pursuers after his defeat by Cromwell's troops at the battle of Worcester, in 1651. This very English name was never likely to be accepted by other astronomers and, sure enough, was politely ignored by sky mappers of other nationalities. Alongside Cor Caroli is a much fainter star called Chara, which is the name of one of the two dogs, Chara and Asterion, represented by Canes Venatici. Along with Castor and Pollux, Chara is one of the very few Greek names among the Arabic.

A number of stars have names that point to some peculiar feature of the star. In Cetus, for example, we find the star Mira. On August 13, 1595, David Fabricius, a Dutch pastor, was looking at the constellation of Cetus and noted a star of magnitude 3 within it—but by October 1 the star had disappeared. Then, in 1603, Johann Bayer saw the star again. He considered it magnitude 4 and listed it as Omicron Ceti. But again it disappeared. Finally, in 1638, the Dutchman Johann Phocylides Holwartha, a professor at Franeker University, started to track the star systematically and discovered that it appeared and disappeared on a regular basis. It was later shown that the star has a cycle of approximately 331 days. Because of its unusual nature, the star was christened "Mira"—the wonderful one—and is now often referred to as Mira Ceti.

Another case is the star Algol—that is, "al-Gol," the Demon—in the constellation Perseus. On a star map, Algol represents the head of the Gorgon Medusa that Perseus holds in his hand. However, the name is apt in another sense in that the star "winks" every three days, growing much fainter over a period of five hours before brightening again, making it the only star visible to the naked eye whose brightness varies so rapidly and obviously. Actually, there are at least three stars in the Algol system, although some astronomers have suggested, probably erroneously, that there may be as many as five. Algol has a larger and much fainter companion that moves in front of it every 2.87 days, thereby cutting off the light of the brighter star. When Algol is hidden (astronomers tend to say "eclipsed," albeit incorrectly because obviously a star cannot have a shadow), we see a third star, intermediate in brightness between the two, making it appear that Algol dims and then grows brighter again. Some have speculated that, because of the name given the star, this variation in luminosity may have been noticed long before the Italian Geminario Montanari reported it in 1699, but there is no proof of this.

Antares is yet another name that calls attention to an idiosyncrasy. At dawn in late March and early April 2002, Mars was close to Antares, as it will be again in late October 2010, just after sunset. Of all the stars in the zodiac that Mars can encounter, only Antares is similar to it in color and brightness, as sky watchers were able to see in 2002 and can look forward to seeing again. The ancients therefore gave it the name Antares, which means "the Rival of Mars," Ares being the Greek god of war, the equivalent of the Roman Mars.

Perhaps the two most curious names, though, belong to the two brightest stars in the constellation Delphinus (the Dolphin). Alpha and Beta Delphini were named Svalocin and Rotanev, respectively, by the astronomer Nicolaus Venator. Some books comment gravely that the source of these names is unknown. But if you think backward for a moment, it will be clear where the names came from. How these names slipped through the net and became official is a mystery.

The Man with the Golden Nose

On December 14, 1546, a son was born to the Danish nobleman Otto Brahe. Christened Tyge, the son was to have a profound influence on modern star maps. His father wanted him to study law, but Tycho, as he started to sign himself after leaving home, wished above all to study astronomy. On his father's death, he duly aborted his legal studies to move to Rostock University and pursue his preferred choice of study. It was there that he lost part of his nose in a duel with a fellow student. Unconcerned, he had a replacement made from gold, silver, and wax that he wore for the rest of his life.

Tycho was a keen observer. His measurements of the planets eventually made it possible for Kepler to prove that the Ptolemaic system of planets rotating around the Earth in perfect circles was incorrect—although Tycho himself proposed a curious system in which the planets circled the Sun and the Sun itself circled the Earth. Using instruments that had no magnifying lenses—he died in 1601, seven years before Galileo demonstrated the capabilities of the telescope—Tycho produced a catalogue that described the position of 777 stars, with a precision far in advance of anything previously achieved. Included in his catalogue was a new constellation, located between Boötes and Virgo, in an area filled with many rather faint stars. This new constellation, which he called Coma Berenices (Berenices's Hair), was the first addition to the forty-eight constellations in Ptolemy's list. Tycho even specified the legend attached to his new addition. When the Egyptian king Ptolemy Euergertes launched his expedition against the Assyrians, his wife, Berenice, swore to cut off her hair and offer it to Venus if her husband returned safe and sound. He did, his wife kept her promise, and Jupiter then placed the shining tresses in the sky as a tribute.

In 1603 Johann Bayer added the southern sky to his *Uranometria*. In so doing, he added eleven new constellations in regions around the South Pole, areas that were far too far south for Ptolemy to have seen. Although Bayer made no attempt to keep to the tradition of attaching legends to constellations (all but one of his additions were named for an animal of some kind), no one has found reason to offer serious criticism of his work, and the additions were rapidly accepted.

Once Tycho had shown the way by making an addition to Ptolemy's list, others besides Johann Bayer flooded into the breach. Each new star map included yet more constellations as astronomers from various countries, apparently on some sort of nationalistic spree, attempted to immortalize their native lands, or in some cases their patrons, by imprinting them on the sky, thereby abandoning the tradition of naming constellations after figures from Greek and Roman legend. Soon the whole situation had turned into a headlong race to "colonize" the sky, especially after 1679, when a slew of celestial cartographers set to work. Between 1679 and 1700, in a flurry of effort to fill in some of the emptier regions of the sky that lay among the traditional constellations, Augustine Royer added five more constellations (two of which were, some 250 years later, accepted into the list of constellations approved by the IAU), Jan Hevelius added eleven (nine accepted), Edmond Halley one (rejected), and John Flamsteed two (both rejected). The wave crested in 1752, when the French astronomer Nicholas Louis de Lacaille added fourteen more constellations to barren areas of the sky. These were the last new constellations later to be generally accepted by the international community and later ratified by the IAU and then included in its definitive list of constellations— although none of them have any compelling reason to exist.

All the same, the trend continued. Between 1776 and 1780 a series of astronomers attempted to add further constellations to the sky, many of them again intended to honor their patrons or their countries. A probably incomplete list indicates fourteen groupings proposed by five different astronomers, all of which were rejected. Readers can decide for themselves the merits of such constellations as Taurus Poniatowski (Poniatowski's Bull), proposed in 1777 by the Polish astronomer Abbé Poczobut of Wilna to honor King Stanislaus Poniatowski, or Sceptrum Brandenburgicum (the Sceptre of Brandenburg), created by the Prussian astronomer Gottfried Kirsch to honor the Brandenburg royal family, or Machina Electrica (the Electric Machine), named by the German astronomer Johann Ehlert Bode around 1775, in honor of the newly created electric generator. The only perhaps regrettable loss is Musca Borealis (the Northern Fly), a small group of stars created by Johann Bode that stood on the back of Aries (the Ram). Despite its official rejection, some astronomers were still referring to this constellation at the end of the nineteenth century. In addition, Poniatowski's Bull still survives as the name of a small asterism in the constellation of Ophiuchus consisting of four stars that form a T-shape. At least at the start, general consensus was the main reason that a new constellation was recognized: if it offended no one, it would be adopted onto star charts by other astronomers.

Putting the Heavens in Order

Astronomers obviously added constellations somewhat willy-nilly, and nowhere was it defined precisely what the borders of the constellations were or which stars belonged to what constellation. Consequently, by the mid-nineteenth century star maps published in different countries showed completely different sets of constellations. Even new planets were initially named in a somewhat random fashion. For many years Uranus, which was discovered in 1781 by William Herschel, was listed in British publications as the "Georgian Star," in honor of Herschel's benefactor, George III. Neptune was known for a time as Leverrier, after the nineteenth-century French astronomer Urbain Jean Joseph Leverrier who, independently of the Englishman John Couch Adams, calculated its position, which led to the planet's discovery in 1846.

The IAU finally appointed a commission to study the problem. The commission made its report in 1928, and the results were published in 1930. Once the recommendations of the commission were ratified, in 1933, an official list of the constellations, their names, and their boundaries was established. The IAU defined the borders of the constellations in terms of meridians and parallels in the sky, which had the curious result that some stars moved from one constellation to another. Sometimes these were stars that had traditionally been shared by two constellations. To take one example, Alpheratz, or Delta Pegasi, was one corner of the well-known Square of Pegasus, but it was also regarded as part of the constellation Andromeda and thus listed as Alpha Andromedae. After the IAU reform, though, Alpheratz was no longer considered part of Pegasus but was instead permanently assigned to Andromeda.

As the international governing body of astronomy, the IAU lays down the rules and recommendations about how new stars and other celestial objects should be named. In most cases these days, the new name consists purely of a catalogue designation, which according to the IAU's guidelines should be descriptive, giving a precise position and defining the type of catalogue. Unfortunately, given the need to catalogue hundreds of millions of stars and galaxies, these modern-day names tend to be rather cumbersome. For instance, with the aid of the 2.5-meter Isaac Newton Telescope in La Palma, I recently had a chance to observe a newly discovered quasar that rejoiced in the name of 87GB073840.5+545138—not something that rolls easily off the tongue. To astronomers, though, these names are logical, informative, and not open to abuse.[5]

Can I Buy a Star?

A number of organizations now offer the public the chance to buy and name a star (at a price) for a friend, a loved one, or even as a present to themselves. These organizations claim the star name will be entered in their registry and will thus, somehow, be official. But the IAU controls the naming of stars, so no name is official without their approval, on top of which most stars, even those of extremely faint magnitudes, have already been catalogued and thus named. In 2003 the United States Naval Observatory published the third edition of their star catalogue, known to astronomers as *USNO B1.0.* This catalogue lists no fewer than one thousand million stars, some of them even fainter than magnitude 20, and gives them each an official name in the form of the IAU-approved catalogue designation.

So, however romantic it may seem to offer your loved one a star as a present, do not be fooled. If you buy a star name from one of these companies, the purchase has no legal validity, and you may very well be buying a star that has already been "sold" to a number of different people. Moreover, that star almost certainly

 What Our Forebears Knew

already has a name, and the only people who will recognize the name you decide to give it are those who have taken your money.

SUGGESTIONS FOR FURTHER READING

Popular Books

Patrick Moore, *History of Astronomy* (London and Sydney: MacDonald & Co., 1983).

> *A wonderful book first published in 1961 as* Astronomy, *later rereleased as* The Story of Astronomy, *and regularly revised since. The volume offers a wealth of detail, spanning legend, biography, history, and science, about the people and events that have shaped the heavens as we know them today.*

Donal Menzel, *Field Guide to the Stars and Planets* (London: Collins, 1975).

> *A compendium of information about the stars and constellations, their history, and the myths that attach to them. An essential book for the serious sky watcher.*

Nathaniel Hawthorne, *Tanglewood Tales* (Boston: Houghton, Mifflin and Company, 1883). Also available online at:
www.pagebypagebooks.com/Nathaniel_Hawthorne/Tanglewood_Tales/

> *A novelist's recounting of many of the most famous legends about the constellations and how they came to be named. Strongly recommended for those who want to learn more about Greek and Roman mythology.*

More Advanced Reading

Mark R. Kidger, I. Pérez-Fournon, and F. Sánchez, eds., *Internet Resources for Professional Astronomy* (Cambridge: Cambridge University Press, 1999).

> *A comprehensive guide to star catalogues and other information about the stars and, in particular, the resources available on the Internet. A valuable reference for the serious amateur or university student.*

On the Internet

The British Museum Compass Facility
www.thebritishmuseum.ac.uk/compass/

> *This site allows one to search for information on, and to locate images of, some three thousand British Museum exhibits, including those that have focused on ancient texts.*

How Did the Stars Get Their Names?   47

The Mythology of the Constellations
www.emufarm.org/~cmbell/myth/myth.html

> *This site offers descriptions of the constellations and their mythology. Click on a constellation name and you receive a detailed account of its mythological origin. Although not every constellation that has mythological connections is included, the site lists a number of the best-known constellations.*

Biography of Ptolemy
www.nineplanets.org/psc/theman.html

> *A short biography of Ptolemy with a series of useful links, from the Nine Planets Web site of the world famous Lunar and Planetary Laboratory at the University of Arizona.*

Life and Work of Ptolemy
http://www-groups.dcs.st-and.ac.uk/~history/Mathematicians/Ptolemy.html

> *An excellent site that surveys Ptolemy's life and work as well as the history of Greek science and mathematics.*

The IAU's Rules and Recommendations for Naming All Stars and Other Astronomical Objects
http://cdsweb.u-strasbg.fr/iau-spec.html

> *The official site that lists the IAU's recommendations on how all celestial objects should be named. These include sources of radiation, given that all objects emit radiation, be it radio waves, visible light, gamma rays, X rays, or any other type of electromagnetic radiation. Conveniently organized and very informative, this site is the place to discover how and why stars and galaxies came to be included in the standard modern catalogues.*

CHAPTER 3

What Was the Christmas Star?

The star of Bethlehem is one of the perennial topics of the press and radio at Christmastime, especially when Venus is visible in the sky. For at least four centuries, historians, biblical scholars, astronomers, and a host of others have explored the nature of the star and have devised theories to explain it. In fact, the star has in all likelihood been the subject of debate since the fourteenth century, and possibly even long before that, in the first few centuries A.D. To paraphrase Winston Churchill's famous comment: "Never in the field of scientific debate has so much been said by so many based on so little."

Millions of words have been written about the star of Bethlehem, but the evidence is so thin and the clues so contradictory that it remains a mystery. Given the paucity of hard facts, even though a number of the proposed explanations of the star have their merits, none is truly convincing. In December 2001 I participated in a three-way debate with two other astronomers, Sir Patrick Moore and David Hughes, about the star of Bethlehem on the BBC's *Sky at Night* program. Each of us defended a popular theory regarding the star. Following the program people were able to vote via the Internet for the theory they found the most convincing. A strong second in the vote was the cover-all option "some other theory," which rather put us in our place.

It is perfectly possible that the star of Bethlehem never existed, in which case there is nothing to explain. Then again, some say it did exist but that it was a miraculous event, not an astronomical phenomenon. Here we will seek a scientific explanation for the star, leaving aside the fact that one may not exist if either of these two alternatives is true.

What the Bible Says

Our story starts with the Bible. In chapter 2 of the Gospel of Matthew we read:

1   In the time of King Herod, after Jesus was born in Bethlehem of Judea, wise men from the East came to Jerusalem,

2   asking, "Where is the child who has been born king of the Jews? For we have seen his star at its rising, and have come to pay him homage."[1]

And Matthew 2:7–10 are as follows:

7   Then Herod secretly called for the wise men and learned from them the exact time when the star had appeared.

8   Then he sent them to Bethlehem, saying, "Go and search diligently for the child; and when you have found him, bring me word so that I may also go and pay him homage."

9   When they had heard the king, they set out; and there, ahead of them, went the star that they had seen at its rising, until it stopped over the place where the child was.

10   When they saw that the star had stopped, they were overwhelmed with joy.

11   On entering the house, they saw the child with Mary his mother; and they knelt down and paid him homage. Then, opening their treasure chests, they offered him gifts of gold, frankincense, and myrrh.

The information in these verses is very scant. The word "star" is mentioned just four times. We learn that the star had been seen sometime in the past at its rising—in other words, low in the east at dawn. We then discover that the star, previously in the east, was now in the south because it "went before them" on the road from Jerusalem to Bethlehem, ten kilometers to the south. Some people interpret verse 9 as implying that the star disappeared between the time the magi set out for Jerusalem and their arrival there but then reappeared to guide them to Bethlehem. This may indicate the star was just a normal star. As the Earth revolves around the Sun, the stars shift their position, so a star will, over several months, move from the east to the south at dawn, and anything other than an exceptionally bright star would seem to disappear if the Moon passed nearby, particularly when it was full.

Two Gospels describe the Nativity, those of Luke and Matthew, although their accounts differ quite considerably. Luke's Gospel comes closer to what we would call historical, but even so it was written as a work of faith. For Luke, the salvation Jesus spoke of was for everyone, no matter what their social status, and his Gospel places great emphasis on the role of women in the life of Jesus. Perhaps for this reason, Luke offers a fairly detailed account of the Nativity, in which Mary plays a prominent part. Indeed, some of the most famous episodes in the story of the Nativity are found in Luke alone, in contrast to the more summary account offered in Matthew. Luke refers to the census decreed by Caesar Augustus and the subsequent journey of Mary and Joseph to Bethlehem to be registered. He also speaks in some detail about the apparition of the angel to the shepherds and of their subsequent homage to the baby Jesus. But at no point does Luke say a word about the magi or the star. In fact, the only New Testament references to the star of Bethlehem appear in the Gospel of Matthew. Matthew mentions the star several times, verbally tracing the arc of its rising and setting above "the place where the child was."

Matthew's image is somewhat troubling. Why is there no mention of the star in the other Gospels if it was as important a feature of the Nativity as Matthew's account suggests? Part of the answer may be that the Gospels were written by different people at different times and intended for different audiences. They were also based on oral tradition—stories about Jesus that were shared among various early Christian sects. Storytellers would recount certain events but not others, either because these were the events they had personally witnessed or because that was how the story had been told to them. Some parts of the story of Jesus appealed to particular audiences and thus were handed down, while others were largely ignored. And, of course, many versions of the story of Jesus were not incorporated into the Bible at all but are instead preserved in the so-called Apocryphal Gospels.

All the same, the lack of common elements between the descriptions of the Nativity provided by Matthew and by Luke is somewhat worrying if we are dealing with historically factual accounts. A common explanation of why Matthew mentions the visit of the magi while Luke refers only to the visit of the shepherds is that, in contrast to the traditional scenes depicted in Christmas cards and nativity plays, these two events were far from simultaneous. Some experts suggest that the visit of the shepherds occurred immediately after the birth of Jesus, whereas the visit of the magi took place weeks, months, or maybe even more than a year later. The debate centers on whether the Greek word translated as "child" in Matthew's account—"Where is the child who has been born king of the Jews?" (2:2), "Go and search diligently for the child" (2:8), "they saw the child with Mary" (2:11)—means "newborn infant" or "toddler." The latter suggests that whereas Luke deals only with events immediately around the time of the Nativity, Matthew covers a wider expanse of time, albeit in less detail.

Matthew appears to have been a Jewish Christian, one who viewed Christianity as a newly evolving religion in transition from orthodox Judaism, not in opposition to it. Hence, Matthew's narrative is inclusive of the Jewish sacred texts, and he quotes repeatedly from the Torah. Matthew also presents the birth of Jesus, as well as certain other of the events in his life, as the fulfillment of the words of Jewish prophets, repeatedly using some variation on the phrase "and so the prophesy was fulfilled." Thus, when Herod inquires of his priests where the Messiah is to be born, they tell him: "In Bethlehem of Judea; for so it has been written by the prophet" (2:5). Similarly, Matthew's account of the flight of Mary, Joseph, and the child Jesus from Bethlehem into Egypt to escape Herod's massacre of the innocents concludes with: "This was to fulfill what had been spoken by the Lord through the prophet, 'Out of Egypt I have called my son'" (2:15). And other such statements—"Then was fulfilled what had been spoken through the prophet Jeremiah" (2:17), "so that what had been spoken through the prophets might be fulfilled" (2:23), and so on—occur throughout his Gospel. Surprisingly, though, Matthew does not cite the prophecy of the soothsayer Balaam, made centuries before the birth of Christ, which is found in Numbers 24:17:

I see him, but not now;
I behold him, but not near—
a star shall come out of Jacob

and a scepter shall rise out of Israel;
it shall crush the borderlands of Moab
and the territory of all the Shethites.

Balaam's words have been interpreted as indicating that a star will appear at the birth of the Messiah, the anointed one of God. If so, one would naturally expect Matthew to point out that the star of the Nativity had appeared just as Balaam had predicted. The absence of Matthew's characteristic "and so the prophecy was fulfilled" has led some scholars to suspect that star never really existed but was merely a rhetorical device Matthew employed to give greater import to the Nativity.

What about other references to the star? Although the star of Bethlehem appears in only one of the four New Testament Gospels, two of the Apocryphal Gospels, James and the First Book of the Infancy of Jesus, also mention the star, although James's account is prone to exaggerate:

> And he questioned the Magi and said to them: "What sign did you see concerning the newborn King?" And the Magi said: "We saw how an indescribably great star shone among these stars and dimmed them, so they no longer shone, and so we knew that a King was born for Israel."[2]

In order to dim the other stars, James's star would have had to be brighter than the Moon, which is obviously impossible. All the same, James does tell us how the star was interpreted by the magi, which indicates that in their eyes it was an event of immense astrological significance. To unravel the mystery of the star further, however, we must set aside the Apocrypha and instead look at how the Western calendar came about.

The Birth of Jesus and the Origin of the Western Calendar

Was Jesus born on December 25? Was the year 2000 the two thousandth anniversary of his birth? The answer in both cases is almost certainly "no." The modern calendar dates from approximately a century after the fall of the Roman Empire. Both the year and the date of Christmas were fixed in 525 A.D. by Dionysius Exiguus, a Scythian monk and church scholar who lived in postimperial Rome.

He chose the name Exiguus, "the Little One," as a mark of humility, to distinguish himself from the more famous Dionysius of Corinth, an influential writer in the early Christian church some two centuries earlier.

The calendar, as it stood, was inextricably linked to the pagan practices of the old Roman Empire, so there was a need to cleanse it of blasphemous references. Dionysius was called upon to construct a new Easter cycle for the church. In freeing the calendar from its Roman roots, Dionysius decided to base his new calendar on the date of the birth of Jesus. First, however, he had to calculate the date of the Nativity. This he did by following precedent and using one of most reliable measurements of time available to him: the reigns of the Roman emperors. Dates in the Roman calendar were based on the number of years after the founding of Rome (*ab urbe condita*, or AUC). By adding the years of the emperors' reigns and working backward, Dionysius was able to fix the date of the birth of Jesus in a particular year AUC.

Dionysius committed two serious errors in his calculations, however. First, he omitted the year 0—for the understandable reason that the Romans did not use zero. Our calendar therefore jumps from 1 B.C. to 1 A.D., without a year 0 in between (and for this reason the third millennium began on January 1, 2001, whatever the politicians may have said). Second and more serious, Dionysius neglected to include an initial four-year period, from 31 to 27 B.C., during which Caesar Augustus reigned under his given name, Gaius Octavius, or Octavian, as he is commonly called. (Julius Caesar had earlier adopted Octavian as his heir, and so, following Caesar's assassination in 44 B.C., Octavian changed his name to Gaius Julius Caesar Octavius. He became known as Caesar Augustus after the Roman senate awarded him the title "Augustus" in 27 B.C.) Given that Augustus was the emperor at the time Jesus was born, this error is not trivial—although it was not until 1605 that a Polish astronomer named Laurentius Suslyga finally pointed out the four-year error in Dionysius's calendar. These two errors totaled five years. Assuming Dionysius made no other mistakes, Jesus would have been born in 5 B.C. or, paradoxically, five years before Christ.

How does this tie in with other information? Contemporary chronicles state that Herod died between the date of an eclipse of the Moon visible from Jericho and the date of Passover. Most experts believe that the eclipse was the partial one seen on the night of March 13–14, 4 B.C., exactly one month before Passover. Jesus was obviously born before Herod's death, which means that his birth must have taken place prior to 4 B.C. This would be consistent with the date of 5 B.C. for the Nativity that we get when we correct Dionysius's two errors. At the same time, Luke offers some confusing clues. In chapter 2 of his Gospel he states:

1   In those days a decree went out from Emperor Augustus that all the world should be registered.
2   This was the first registration and was taken while Quirinius was governor of Syria.
3   All went to their own towns to be registered.

Caesar Augustus reigned from 31 B.C. to 14 A.D., so the reference to him in verse 1 does not help much. The second verse, though, contains two highly valuable clues: the observation that this was the first census and the name of the governor of Syria at the time of the Nativity. Caesar Augustus decreed the taking of three censuses for Roman citizens: in 28 B.C., in 8 B.C., and finally in 14 A.D. The date of the 8 B.C. census is known because the decree ordering it was found some years ago at Ankara, in Turkey. It is usually assumed to be Luke's census, despite the fact that Joseph was not a Roman citizen and so would not have been required to respond to the registration order. Moreover, this census is a little early to be consistent with a date of 5 B.C. for the birth of Jesus. Of course, writing almost a hundred years later, Luke might have committed an error. He might have been referring to a decree ordering those available for military service to report for registration or even a demand that all conquered people register an oath of allegiance to the emperor.

What about Quirinius? Here we appear to have a major problem. Quirinius did not become governor of Syria until 6 A.D., ten years after the presumed date of Herod's death. However, Quirinius was the emperor's legate in Syria between 6 and 5 B.C., under the governership of Saturnius, and Luke may simply have confused the two posts. If this hypothesis is correct, the dates of Quirinius's legateship do agree well with the death of Herod and with the Dionysian calendar. The best guess, then, is that the Nativity took place in 5 B.C., although 6 B.C. is also a possibility given that, in seeking to kill Jesus, Herod ordered the slaughter of all children less than two years old. We are not certain exactly how long before his death this order was made, although it is generally agreed that it was one of his last acts as king.

What about the date of Christmas? Again, there is little, if any, historical basis for the date we observe. December 25 has been celebrated as the date of Jesus's birth at least since 336 A.D., and possibly much earlier, since 336 is merely the first time the date is mentioned in the records surviving to us. The date is taken from the ancient pagan festival of Sol Invictus, marking the shortest day of the year, and represents a christianization of long-established and highly popular festivity. In fact, our modern Christmas seems to be closer to the traditions of the pagan Sol Invictus, which included the giving and receiving of gifts, the decoration of houses with green branches, feasting, and a general atmosphere of revelry, than to the historical circumstances of the Nativity. Rather than attempt to abolish a traditional holiday with a millennium of history behind it—it had been celebrated by Celtic peoples well before the founding of Rome and was subsequently adopted by the Romans themselves—the early Christian church adopted and adapted it. In other words, the date of Christmas Day was fixed more or less arbitrarily.

Some fifty years ago, Werner Keller pointed out that the Nativity almost certainly did not take place in winter. Luke states that:

> 8   In that region there were shepherds living in the fields, keeping watch over their flock by night.

As Keller observed, December in Palestine is typically characterized by heavy rainfall, and over the entire winter period there are, by turns, rainstorms, frost, and occasionally even heavy snowfalls. Bethlehem is at an altitude of 765 meters above sea level, high enough that snow showers often sprinkle the hills, burying the forage. Flocks would usually be under shelter, and shepherds would certainly not be spending their nights in the fields. Rather, shepherds would be out on the hills at lambing time, in March, April, and possibly early May.[3]

Thus the most likely date we have for the Nativity is the spring of 5 B.C. Now we need to look at what the key players in the story of the Nativity might have to tell us.

Who Were the Magi?

Investigations into the star of Bethlehem generally pay surprisingly little attention to the magi and their origins, and to the clues that they offer. However, if

What Our Forebears Knew

we are to understand the star, we must try to determine who the magi might have been and consider what kind of celestial sign would have been significant to them.

Matthew is not helpful referring only to "wise men from the East." He uses the plural, so at least we know that there were more than one. In the oldest representations of the Nativity the number of magi varies widely: sometimes two, sometimes three, other times four, or even twelve. To add to the uncertainty, in place of "we have seen his star at its rising," older translations of Matthew read "we have seen his Star in the East," making it somewhat unclear whether it was the magi or the star, or both, that were in the east. The word *magi* comes from the Greek for "magician," which hints that they possessed special powers. It is very unlikely the magi were kings: Herod does not treat them as visiting royalty, and it is only in the sixth century A.D. that we first find mention of them as kings. Most authorities now accept that the magi were astrologers who were charged with interpreting the signs observed in the sky.[4] In that case, the star would not necessarily have had to be a particularly spectacular object in order to be of major astrological import to those who "knew" how to interpret it.

Although one reads in many books that the magi were Babylonians, little evidence exists to support this idea, which in fact seems increasingly unlikely with each passing year. We are aware that, although they were far behind the Chinese, the Babylonians did have a knowledge of astronomy, for a number of their observations have survived. For example, the only record of the appearance of Halley's comet in 163 B.C. comes from Babylon, and, as we have seen, so do many of the modern constellations, including those of the zodiac. The idea that the magi were Babylonian, however, appears to derive solely from the historical preeminence of Babylon in the region.

What of the alternatives? During the first centuries A.D., tradition had it that the magi were Arabian, in accordance with predictions in the Scriptures, but there seems to be little else to link them with Arabians. If the magi were Arabian, the romantic idea of a journey across the desert on camelback to Jerusalem would need to be put to rest. In all likelihood, Arabian magi would have sailed around the coast and disembarked at the port of Erion-Geber at the tip of the Red Sea, riding from there a short distance up the King's Highway to Jerusalem. This journey would have been rapid and relatively comfortable.

More plausible, though, is the Persian origin of the magi. Although the evidence is tantalizing rather than conclusive, it is increasingly compelling. In the

Apocryphal Gospel of the Infancy we read that the magi came to Jerusalem "according to the prediction of Zoroaster":

> And it came to pass, when the Lord Jesus was born in Bethlehem, a city of Judea in the time of Herod the King: the wise men came from the East to Jerusalem, according to the prophecy of Zoradascht [Zoroaster], and brought with them offerings, namely, gold, frankincense, and myrrh, and worshipped him, and offered to him their gifts. (3.2)

In biblical times the magi were a caste of priests, specifically Median priests of Zoroastrianism, the principal religion of ancient Persia. Founded by Zoroaster around 1000 B.C., the religion still survives today in some parts of Iran. Zoroaster is said to have been born in the town of Urmiyah, in what was historically Persia but is today Iran. He preached that there was one god, Ahura Mazda ("Wise Lord"), who was the force of good in the world and was opposed by the forces of evil. Some Zoroastrian writings appear to be messianic. They predict that a son of Zoroaster will be born many years after his father's death, to a virgin who bathes in a lake in which Zoroaster's semen has been preserved. The son of Zoroaster will raise the dead and crush the forces of evil. This prediction has been interpreted by some as prophesying the birth of Jesus.

When, in the late thirteenth century, Marco Polo crossed the region of what was then northern Persia on his way to China, he was told by the proud citizens of Säveh, a small town in Markazi province, about 150 kilometers southwest of Tehran, that their town had been the point of origin of the magi. Other towns in this region make the same claim, which suggests that it may indeed have some historical basis. In addition, the earliest carvings of the Nativity (although from several centuries after the event) depict the magi in Persian dress, wearing belted tunics, full sleeves, trousers, and Phrygian caps rather than the traditional robes that we see in most representations of the Nativity. There is even a legend that a carving of the magi saved the church of the Nativity at Ravenna, on the Adriatic coast of northern Italy, from the ravaging Persian hordes in 614 A.D. When the invading army saw the Persian figures on the walls inside the church, they spared the building from pillage and torching.

The theory that the magi were from Persia does face one difficulty, namely, that there is no known tradition of Persian astronomy. But inasmuch as there are few surviving historical records of any kind from Persia, this does not necessarily mean that the Persians had no interest in astronomy. The most likely conclusion thus seems to be that the magi were Persian and may have been astrologer-priests awaiting the birth of Zoroaster's son (as they would have taken the Jewish Messiah to be). Their star would have been a sign in the sky foretelling the birth of a king and would therefore have been an event of overwhelming astrological significance.

Amazing and Unlikely Theories

Many imaginative theories have been proposed to explain the star of Bethlehem. The planets Venus and Uranus, ball lightning and an auroral display, meteors and Halley's comet, a supernova, and an occultation are some of the more unlikely candidates. We will examine these only briefly before moving on to more promising ideas.

Venus

When Venus, the brightest of all planets, is visible at Christmastime, as it is most years, people often wonder whether it might be the star of Bethlehem. But Venus was a familiar celestial object for many centuries prior to the Nativity, having first been recorded in 1600 b.c. by the Babylonians. No astronomer or astrologer would be fooled into thinking that this planet was a new or unusual star.

Uranus

In 9 b.c. Uranus passed close to Saturn, and then in April 6 b.c. it passed close to Venus. These twin astronomical events have led some to suggest that it was Uranus the magi initially noticed (some eighteen hundred years before William Herschel was finally credited with its discovery) but then lost sight of, only to see again when the planet moved into conjunction with Venus, some three years later. If this theory is correct, the magi must have been supremely patient! To the naked eye, however, Uranus is only dimly visible and thus would make a poor star. It is roughly as bright as the asteroid Vesta, one of the largest of these untold millions of rocks that orbit the Sun between Mars and Jupiter. If the magi had discovered Uranus, surely they would have noticed Vesta as well since it moves more quickly among the stars and is therefore more conspicuous. More important, the conjunction in 9 B.C. occurred so close to the Sun in the sky that it would have been extremely difficult to observe.

Ball Lightning

Ball lightning is a rare atmospheric phenomenon, yet to be fully understood, that consists of globes of light that seem to hover in the air or move along solid objects. The chief support for the ball-lightning theory is therefore that it would allow the "star" to lead the magi and then to hover and stop. Ball lightning is short-lived, though, usually lasting for no more than a few minutes. It is just not plausible that the magi would see it, travel for weeks, and then find it again at just the right moment.

A Meteor

We are all familiar with the sight of meteors, or shooting stars, as they are popularly known, flashing across the sky. The theory that the star of Bethlehem could have been a meteor has been popularized over the years by Sir Patrick Moore. This theory suffers from similar problems, though. Ordinary meteors are common and so would be familiar to the magi, but they are normally visible for no longer than a second. A meteor, or stream of meteors, crossing the sky might indicate a direction in which to travel, but to impress the magi the meteor shower would have to be exceptionally bright, and it would disappear far too fast to allow them to react.

The Aurora Borealis

The aurora borealis, or northern lights, would certainly be a striking sign in the heavens for the magi, especially given that the lights are very seldom seen as far south as the Middle East. Again, though, the auroral display is a short-term phenomenon. Furthermore, given that the aurora borealis appears in the north, it would not be able to lead the magi south to Bethlehem.

A Near-Earth Asteroid

Another very unlikely idea. This theory suggests that a small rock from outer space passed so close to the Earth that it was visible to the magi as a bright, moving star and that they then followed it as it traveled across the sky. As far as we know, however, never has a near-Earth asteroid (so called by astronomers because their orbits differ from those of most asteroids such that NEAs can pass close to the Earth) been bright enough to be seen without a telescope.[5] And even supposing one were that bright, it would be visible at most for a few hours, disappearing so fast that it could not have kept company with the magi for their entire journey, however fast their camels.

Halley's Comet

The theory that the star of the magi was in fact Halley's comet was popular in the eighteenth century, when astronomers first calculated that the comet must have appeared close to the date of the Nativity. (As we will learn in the following chapter, the history of this famous visitor to Earth is closely linked to the story of the star of Bethlehem.) We know now that the comet was observed by the Chinese in the autumn of 12 B.C. As it seems impossible that Jesus was born before 7 B.C., the interval between the appearance of the comet and the magis' trip to Bethlehem would have been at least five years, and probably more like seven. But if the star of Bethlehem was indeed Halley's comet, then the magi took seven years to respond to it and make what is usually assumed to be a journey of no more than two months—so they were obviously in no hurry!

Some Other Comet

Those who have been fortunate enough to see a bright comet—perhaps Comet Hale-Bopp, in 1997—will know what an impressive sight one is, with its tail trailing behind it across the sky. However, the idea that the star of Bethlehem was a comet is implausible. Comets have long been associated with death and disasters, and thus it is highly unlikely that the magi would interpret a comet as signaling the birth of the Messiah. Moreover, apart from the observation of Halley's comet in 12 B.C., there is no record of a bright comet having appeared at any time in the twenty years surrounding the Nativity.

A Supernova

This is an idea popularized in the 1950s by Arthur C. Clarke in various of his stories and articles. A supernova is a stellar death—an old star that blows itself apart in a giant nuclear conflagration. Supernovas may be so bright that they can be seen even during the day. But from Chinese chronicles that have since been studied, we now know that the supernova closest to the date of the Nativity occurred in 185 A.D.

An Occultation

An occultation occurs when a planet or a star passes behind the Moon or some other body and its light is hidden briefly until it emerges from the other side. In the 1990s it was suggested that the star might actually have been the occultation of Jupiter by the Moon on April 17, 6 B.C., possibly in combination with another a month earlier. Unfortunately, the March occultation took place in broad daylight, during the early afternoon, while the April occultation occurred extremely low in the sky, a few minutes after sunset. Both would therefore have been difficult to observe even with a telescope, an item the magi did not have. Moreover, astrologically speaking, the disappearance of the king of planets behind the Moon would seem to suggest the death of a king rather than a royal birth.

Planetary Conjunctions

In 1968 Donald Sinnott, now an editor of the popular magazine *Sky and Telescope*, suggested a new and extremely well-researched idea, namely, that the star of Bethlehem might have been a conjunction of two or more planets. A conjunction takes place when two planets pass nearby each other on a north-south line. When bright planets such as Venus, Jupiter, and Mars are involved, the configuration can be extraordinarily spectacular. Moreover, depending on which planets appear together in the sky and in which constellation this event occurs, astrologers might well interpret the occultation as significant.

What Our Forebears Knew

Sinnott searched through the period from 12 B.C. to 7 A.D. for noteworthy planetary configurations and found more than two hundred conjunctions involving the five brightest planets and twenty groupings of three or more planets. Most of these occurred close to the Sun in the sky or were otherwise not particularly brilliant, but a few stood out. The two most remarkable involved Jupiter and Venus and took place in Leo in August of 3 B.C. and on June 17, 2 B.C. The second of these was an outstanding event. As seen from Babylon, the two planets would already have been quite close at sunset but over the next two hours would have continued to approach each other until, when the pair set, they would appear to fuse into one object. Calculations have further shown that a few hours later, the gibbous disk of Venus, 80 percent illuminated, would have passed in front of Jupiter and partially occulted it. This is an extremely unusual phenomenon indeed, although the magi could not have known about it since the occultation occurred after the two planets had set in the sky. Even so, this conjunction was certainly a very special event and thus a worthy candidate for the star of Bethlehem. Sinnott suggested that the magi might well have associated this spectacular conjunction, which took place in Leo and involved the king of the planets, with the birth of the "Lion of Judah."

Sinnott's theory has great merit, but it meets with two problems. First, it is now generally accepted that the Nativity occurred prior to 4 B.C., whereas the two pivotal conjunctions in Sinnott's theory occurred in 3 B.C. and in 2 B.C. In order for Sinnott's theory to be plausible, the eclipse that marked the death of Herod would have had to be that of January 9, 1 B.C.—the next eclipse of the Moon to be visible from Jerusalem—rather than the one in March of 4 B.C. In other words, Herod would have died some three years later than is generally thought, and the Nativity could similarly have taken place several years later. But this leads to new problems with Quirinius and with Dionysius's chronology, both of which can be reconciled with a date of 5 B.C. for the Nativity but most certainly not with a date such as 2 B.C. Second, the conjunction would have been seen in the west at sunset and would have lasted just a few nights, with the most spectacular configuration confined to a single night. As a celestial signpost, it could have led the magi west to Jerusalem but never south from Jerusalem to Bethlehem.

More recently, the British astronomer and astrohistorian David Hughes called renewed attention to a less visually spectacular conjunction that Sinnott had rejected. In 7 B.C. a triple conjunction of Jupiter and Saturn took place in Pisces, a constellation traditionally associated with the Jews. Triple conjunctions occur when two planets pass close to each other three times in the space of a few months, whereas an ordinary conjunction consists of just a single, brief encounter. On May 29, September 29, and December 4 of 7 B.C. Jupiter and Saturn approached to within twice the diameter of the Moon before separating again. Triple conjunctions are rare (there have been only eighteen since 1000 B.C., although curiously there were three in Pisces between 980 B.C. and 7 B.C.).[6] Such a conjunction, while not visually very striking, could be rich in astrological meaning, but it would probably not have been noticed by Herod and his courtiers, which would explain their ignorance of the star.

Hughes further suggests that the Nativity occurred on September 15, 7 B.C. This was the day that Jupiter and Saturn had their achronical rising, that is, they rose together in the east at the same time the Sun set in the west. In Babylonian astrology Jupiter was the star of Marduk, the supreme god of Babylon. Saturn was the star of the king, the earthly representative of the god, and was called Kaiwanu, "the Steady One." Pisces was associated with Ea, the god of wisdom, life, and creation, and was the last sign in the zodiac and hence the last constellation the Sun passed through each year. The conjunction of Jupiter and Saturn in Pisces accordingly portended two things: the end of the old world order and the birth of a new king chosen by God.

We know that the Babylonians observed the conjunction because a tablet in the British Museum (BM 34429) records it, albeit in rather perfunctory terms: "Month VII, the 1st of which will follow the 30th of the previous month. Jupiter and Saturn in Pisces, Venus in Scorpio, Mars in Sagittarius. On the 2nd, equinox." There is little in these words, which describe October ("Month VII") of 7 B.C., to suggest that the Babylonians found the conjunction in any way an important event. However, there is no doubt in my mind that, whoever the magi were, this triple conjunction was a key event. It told the magi *what* to expect—a great event in Judea—although probably not *when* to expect it.

The Chinese Star

For many years now researchers have pored over the Chinese chronicles that refer to the last two decades B.C. and first decade A.D. in the hope of finding a reference to the star. The Chinese recorded everything assiduously—novas, super-

novas, comets, eclipses, meteor showers, auroras, and even sunspots—over nearly four millennia, first on bones (oracle bones) and later in book form. The observations, often written with ink on silk and then bound into books, were often extremely detailed. In the case of a comet, for example, the Chinese traced its movement in the sky and made note of its shape, the length of its tail, and even its color. However, for some years after the original publication, in 1962, of the catalogue of Chinese astronomical observations, the fact that four events were recorded between 12 B.C. and 1 B.C. was a source of some confusion. It has now been established that the first, in September of 12 B.C., was Halley's comet. The second, in 10 B.C., was soon shown to be a "ghost event" (that is, an error—usually the result of mistakes in later transcriptions of the observations) and thus never existed. But what of the objects seen in the spring of 5 B.C. and the spring of 4 B.C. in the same region of the sky? For many years it was thought that the existence of these two events in such a short span of time would make it impossible to decide which might be *the* star.

In a volume called the *Ch'ien-han-shu*, we find the following reference:

> In the second year of the period of Ch'ien-p'ing, second month, a *hui-hsing* appeared in Ch'ien-niu for more than seventy days.

The second year of the reign of Ch'ien-p'ing was 5 B.C., and the second month (during which the observation was made) ran from March 10 to April 7. A *hui-hsing* is a broom star—a star with a tail, usually a comet—and Ch'ien-niu is the Chinese constellation that included Alpha and Beta Capricornii. So the translation is:

> During the interval between March 10 and April 7 of 5 B.C., a comet with a tail appeared close to Alpha and Beta Capricornii and remained visible for more than seventy days.

Curiously, though, the chronicle states that the object remained fixed in the same place in the sky for more than two months, which means that it was almost certainly not a comet.

As for the object that appeared in 4 B.C., in a twelfth-century Korean chronicle called the *Samguk Sagi*, or the *History of Three Kingdoms*, we find this reference:

> Year 54 of Hyokkose Wang, second month, [day] *chi-yu*, a *po-hsing* appeared in Ho-Ku.

Ho-Ku is the Chinese constellation, or asterism, that includes Altair and the various stars in the southern section of the constellation Aquila (the Eagle). A *po-hsing* is a bushy star: an extremely bright star with rays, or a comet without a tail. The statement in the *Samguk Sagi* poses a problem, however, because the day called *chi-yu* did not exist in the second month of the year. It is as if the chronicler had written that the star had appeared on February 30. Given that the *Samguk Sagi*, although a Korean history, was written in classical Chinese, one way of resolving the problem is to suppose that *chi-yu* really should be *i-yu*, a Chinese character written in an almost identical fashion and easily confused with *chi-yu*. If so, we should read:

> Year 54 of Hyokkose Wang, second month, [day] *i-yu*, a *po-hsing* appeared in Ho-Ku.

Day *i-yu* of the second month of year 54 of Hyokkose Wang was March 31, 4 B.C., which gives us:

> On March 31 of 4 B.C., a bushy star appeared close to Altair.

But even if the confusion of *chi-yu* and *i-yu* correctly explains the error in the day of the month, in 1977 a group of three British astrohistorians, David Clarke, John Parkinson, and Richard Stephenson, called into question the reliability of the year given in the Korean text. As they pointed out, the southern part of Aquila (which, like Altair, is part of Ho-Ku) borders on northern Capricornus, and thus the positions given in the two chronicles are located in the same, very small area of sky. It is difficult to imagine, however, that two such similar objects would appear in the same region of the sky in the same month of two consecutive years. Rather, the Chinese *hui-hsing* and the Korean *po-hsing* were almost certainly the same object, and in all likelihood the year, as well as the day, given in the Korean chronicle was wrong. If so, then the object must have appeared in 5 B.C., during the month of March. Lo and behold—exactly the same date as the most probable one for the Nativity.

The problem with this explanation is the description of the star in the Chinese text as a *hui-hsing* (a broom star, typically a comet with a tail) and in the Korean as a *po-hsing* (a bushy star, or a comet without a tail). The star of Bethlehem was most certainly not a comet—historically, and even in modern times, viewed by both astrologers and ordinary people as a harbinger of disaster. The Chinese chronicles do not state that the star moved. Rather, they appear to suggest that it remained fixed in place for over two months, which is revealing, as it suggests that the designation for a comet, *hui-hsing*, was used wrongly. Occasionally the Chinese chronicles did in fact refer to an object as a *hui-hsing*, even though we now know that it was a star. This happened, for example, in the case of Tycho's supernova of 1572, which the Chinese chronicles variously labeled a *hui-hsing* and a *po-hsing*. (Curiously, some European astronomers also described Tycho's supernova as a comet.) In other words, by calling the object they observed a *hui-hsing*, all the Chinese may have been saying is that it was unusually bright.

The position of the *hui-hsing* (or *po-hsing*) seen at the end of March in 5 B.C. must have been close to Theta Aquilae, a moderately bright star of magnitude 3, found just a few degrees south of the brilliant star Altair. The fact that this star lies within the Milky Way suggests that the *hui-hsing* may have been a nova, or a new star, because these occur within the band of the Milky Way, whereas comets can appear anywhere in the sky. Actually, a nova is an ancient pair of stars—one "dead," or collapsed, dwarf star and a second that is approaching the end of its life. When these two are close to each other, the gravitational force of the dwarf star sucks material from the larger star onto the surface of the dwarf star. This material builds up until it becomes so hot that it explodes in a violent thermonuclear reaction, blasting the fallen gas off the surface of the dwarf star. This we see from Earth as a sudden brightening of what was previously a faint star. For a few days or weeks the nova may increase in brightness a millionfold, or even more, before finally fading.

The evidence thus suggests that the *hui-hsing* observed in 5 B.C. was a bright nova. In the 1960s the Russian astronomer Boris Kukarkin pointed out that even though the star had been visible for quite some time—over seventy days—its disappearance coincided with the arrival of the monsoon season in China. It is quite likely, then, that the Chinese astronomers stopped observing it not because it had grown too faint to see but rather because of bad weather. The nova would have appeared quite low in the east at dawn. One night nothing would have been visible, and the next night a bright star would suddenly have shone in the predawn sky. Like all stars, though, it would rise four minutes earlier each day and gradually move higher in the sky at dawn. Suppose the magi saw this star in the east at dawn and then took two months to reach Jerusalem—time enough to prepare for the journey and then make the long and difficult trek from Persia. After two months the star would now be due south at dawn. In other words, it would have pointed them south toward Bethlehem on the road from Jerusalem, appearing to go before them and then hang over the town of Bethlehem, just as Matthew tells us the star did.

What Was the Star?

As an astronomical event, then, the star must have been the nova seen by the Chinese in March of 5 B.C. In my view, however, the magi would have had no reason to associate this nova with the Nativity had they not had a previous sign to put them on the alert. For if the mystery could be explained by a single event, we would by now have found a convincing answer. There is good reason to believe that the triple conjunction of Jupiter and Saturn in Pisces that occurred in 7 B.C. was an astrological warning to the magi that great events would soon happen in Judea.

However, consider that on April 24, 126 B.C., a more impressive conjunction of the two planets had taken place, again in Pisces, that coincided with a spectacular massing of Mercury, Venus, Jupiter, Saturn, and the Moon in the dawn sky. This was the last of a series of six conjunctions and massings that took place over three months, all involving Jupiter, including a particularly close conjunction with Venus on April 4. The conjunction of 126 B.C. would presumably have conveyed a similar or even more urgent message than that of 7 B.C. What was the difference for the magi? Why did the more spectacular conjunction not rouse them to action? The key is probably the mentality of the times. The magi were a patient group, not easily excited. They read the signs in the conjunction and then waited for the definitive sign, a new and different one, that would tell them of the birth of the Messiah. The magi were alerted to the imminent Nativity by the triple conjunction in 7 B.C. Their horoscopes told them to expect the birth of a king of the Jews. They did not know when or where he was to be born, and so they awaited further signs to instruct them. Possibly an additional sign did appear to them, for we know that in February of 6 B.C., a few months after the triple conjunction ended, a planetary massing took place in Pisces, when Mars joined Jupiter and Saturn in an 8° radius. This new event in the heavens, which was recorded by the Babylonians on the same tablet that reported the triple conjunction, might well

What Our Forebears Knew

have confirmed their expectations. Once again the planets gathered for an astrologically significant event in the constellation of Pisces. This time, in addition to Jupiter, the king of the planets, Mars—the bringer of war—was added to the group, suggesting that the messianic prediction of a new king who would liberate the Jewish people was about to be fulfilled. Finally, after a further delay of a few months, in March of 5 B.C. a nova blazed out in the dawn sky. After the centuries during which their predecessors had waited patiently, weathering many disappointments, it was the birth of a new star in the sky that told the magi that the king had at last been born. This was the celestial signal that told them to make their plans and prepare for their journey. Without this further astrological signal no doubt the magi would simply have gone on waiting until the right combination of signs finally occurred—one that would have told them what was going to happen and where it would happen, and, most important, would advise them of the moment when it had happened so that they could start out on their journey.

But will we ever solve the mystery of the star of Bethlehem once and for all? After two thousand years, is it possible that some piece of evidence will emerge that settles the question? It seems unlikely, although if we could fix the date of the Nativity with greater precision we would be able to narrow down our possibilities still further. Even so, we will probably never know for certain which of the celestial events that took place around the time that Jesus was born was the star. The fact that no single event has yet offered a convincing explanation is perhaps a clue that no one event *can* explain the star. If there is a simple explanation, though, we are probably as close to it as science can get us.

SUGGESTIONS FOR FURTHER READING

Popular Articles and Books

R. W. Sinnott, "Thoughts on the Star of Bethlehem," *Sky and Telescope*, December 1968, p. 384.

> *A classic study. Although events have caught up with it to some degree, in many senses it remains one of the very best short articles ever written on the subject.*

Arthur C. Clarke, "The Star of the Magi" (1954), in *Report on Planet Three and Other Speculations* (London: Corgi Books, 1975).

> *A very worthwhile article, albeit written before the Chinese texts had been translated and thus quite dated in its solution to the mystery—but still extremely readable.*

Arthur C. Clarke, "The Star" (1955), in *The Other Side of the Sky* (London: Victor Gollancz, 1987).

> *This is a fictional counterpart of the previous article. Among other things, the story explores the conflict between scientific explanations and religion. A good read, with a wonderfully unexpected ending.*

Mark Kidger, *The Star of Bethlehem: An Astronomer's View* (Princeton: Princeton University Press, 1999).

> *A volume that offers a thorough review of all the evidence we currently have and looks closely at the assortment of theories that have been advanced to explain the star.*

More Advanced Reading

D. W. Hughes, "The Star of Bethlehem," *Nature* 264 (1976): 513–18.

> *A serious and detailed treatment of the star in which Hughes uncovers and corrects various historical errors in interpretation both of the Bible and of the star itself.*

D. H. Clark, H. Parkinson, and F. R. Stephenson, "An Astronomical Re-appraisal of the Star of Bethlehem—A Nova in 5 B.C.," *Quarterly Journal of the Royal Astronomical Society* 18, no. 4 (1977): 443–49.

> *Probably the best treatment to date of the Chinese texts that may contain references to the star of Bethlehem. A nicely written and persuasive article that provides definitive answers to some of the many questions about the star.*

C. J. Humphreys, "The Star of Bethlehem—a Comet in 5 B.C.—and the Date of the Birth of Christ," *Quarterly Journal of the Royal Astronomical Society* 32, no. 4 (1991): 389–407.

> *While its conclusions are somewhat controversial, this article is meticulously researched in terms of historical documentation and records and is, in this sense, without doubt the most comprehensive work available.*

On the Internet

The *Sky at Night* Debate
www.bbc.co.uk/science/space/spaceguide/skyatnight/proginfo.shtml

> *The televised version, broadcast on the BBC in December 2001, of the debate about the nature of the star of Bethlehem that Sir Patrick Moore (the program's host), David Hughes, and I undertook.*

The Nativity Pages
http://ourworld.compuserve.com/homepages/p_greetham/wisemen/home.html

> *The Reverend Phillip Greetham is an English Methodist minister living in Essex, who has wide-ranging interests that include Christianity and astronomy. His Internet site on the star of Bethlehem is by far the best I have seen and presents an enormous amount of historical and biblical information from a Christian, but also scientific, viewpoint, not marred by strong prejudices or preconceived opinions. I strongly urge readers to take a look at this site.*

Christmas Symbols
http://ww2.netnitco.net/users/legend01/xmas97.htm

> *Information on Christmas and Christmas traditions, including the role of the magi.*

CHAPTER 4

How Do We Know When Comet Halley Was Seen?

Halley's comet is one of the key players in the star of Bethlehem story. Knowing exactly when the comet was visible during the period surrounding the Nativity is vital to any explanation of the star. To some, the claim that we can know such things seems magical at best and outright arrogant at worst. If we can't forecast next week's weather accurately, how can we possibly say what Comet Halley did more than two thousand years ago? Are we really so certain, for example, that it was observed in 12 B.C., when, if we jump back at 76-year intervals from its most recent appearance, in 1986, we would expect it to have been seen around 10 A.D.? It turns out that "as regular as Halley's comet" does not mean quite what we intend. How we have tracked Halley's comet through the centuries is another fascinating tale, one that reveals much about how astronomers today differ from their counterparts of old.[1]

Comets provide one of the few ways that a living person can be immortalized in the sky, for they are usually named after the person who discovered them. Comet discovery is accordingly a highly competitive business, with considerable prestige attached to such discoveries. In the past, medals or cash prizes were sometimes conferred on those who first noticed a particular comet. To encourage amateur astronomers to search for comets this tradition was reinstituted in 1999 in the form of the Edgar Wilson Award, an annual prize of $22,000 to be shared among all amateur discoverers of comets each year. Well before that, however, the competition had already become extraordinarily intense at times. In the 1970s, one comet was discovered by no fewer than five Japanese astronomers within fifteen minutes of one another. The naming of comets is now governed by strict rules, which have ended much of the controversy that was once part and parcel of the rivalries involved. Even so, these rules are occasionally put to the test, as happened with one discovery in the year 2000, when various astronomers claimed credit for the discovery, and a bitter dispute ensued. In the event of a disagreement, a committee of the International Astronomical Union, the Small Bodies Naming Committee, convenes and issues a definitive ruling.

When a new comet appears, the Central Bureau for Astronomical Telegrams (CBAT)—the office of the International Astronomical Union that deals with astronomical discoveries, located in Cambridge, Massachusetts—must be advised of the discovery. Generally the first person to notify CBAT is rewarded with seeing his or her name attached to the comet. This person may not be the first to observe the comet, however. A recent example is the famous Comet Hale-Bopp. Alan Hale actually discovered this comet on July 23, 1995, forty-five minutes after another amateur astronomer, Thomas Bopp, spotted it. But Bopp was observing the sky from deep in the New Mexican desert, and it took him more than three hours to find a way of sending a telegram to the CBAT early on a Saturday morning. Inasmuch as both astronomers discovered the comet independently on the same morning, they were both classed as discoverers, but because Hale's announcement arrived first, he received precedence. Tom Bopp, a good friend of mine, is, I believe, amazingly philosophical about the whole affair and is quite content that his comet is called Hale-Bopp instead of Bopp-Hale.

Up to the early 1990s the IAU allowed as many as three independent discoverers to be recognized. Among other things, this produced some rather cumbersome comet names, at least for Western tongues. Possibly the most long-winded name in recent history is Comet Nishikawa-Takamizawa-Tago, an object independently discovered by four Japanese astronomers in 1987, of whom the first three were recognized. (Shigeo Mitsuma missed being acknowledged as a discoverer by just four minutes—probably just as well, since that would have given us "Comet Nishikawa-Takamizawa-Tago-Mitsuma"!) Decades earlier, one comet wound up with four names when, between 1818 and 1928, four astronomers discovered the same object at different times. Fortunately, Comet Pons-Coggia-Winnecke-Forbes was later renamed Comet Crommelin, after the astronomer, Andrew Crommelin, who analyzed its orbit and proved that the three comets discovered in 1818, 1872, and 1928 were one and the same body, which has a period of 27 years. Since the early 1990s, though, no matter how many people discover a comet more or less simultaneously, only two names are accepted by the IAU. This rule has led to some hard luck stories when, under the old regulations, a third person's name would have been included. The beautiful Comet Ikeya-Zhang, seen in March and April 2002, would have been named Ikeya-Zhang-Raymundo, the last for the Brazilian astronomer who, unaware that the comet had already been reported, discovered it some twelve hours later. Although Paolo Raymundo was accepted as an independent discoverer, the comet does not bear his name. An earlier, rather distinctive, test case was the series of comets discovered in the early 1990s by the husband-and-wife team of Carolyn and Eugene Shoemaker, who often worked in collaboration with David Levy, an amateur astronomer turned professional. In this instance, common sense prevailed, and rather than being called "Shoemaker-Shoemaker-Levy," the comets were simply known as "Shoemaker-Levy."

Recently many comets have been discovered by accident by automated telescopes that scan the heavens in search of asteroids that might one day pose a hazard to the Earth. (For more about the threat posed by asteroids, see chapter 9.) In most cases, the telescope automatically makes note of the object and measures its position. This information is then transmitted to the international clearinghouse for asteroid observations, the Minor Planet Center, housed at CBAT, without benefit of a human intermediary who might notice that the object is actually a comet. This happens especially with the LINEAR (LIncoln Near-Earth Asteroid Research) Telescope in New Mexico, a highly sophisticated telescope that scans the night sky automatically, searching for asteroids, which now has more than a hundred comets to its name. These are named after the telescope, not for individual astronomers.[2] Almost always these comets are first reported as asteroids, but then other observers, often amateur astronomers, identify them as comets. In some cases, in the course of routine observations, an astronomer at some other telescope notices that the object previously reported as an asteroid by the LINEAR

What Our Forebears Knew

Stonehenge

The heelstone

Distant views of the ditch and bank around Stonehenge

The Venus tablet, which describes observations of the planet Venus
made during the reign of King Ammisaduqa

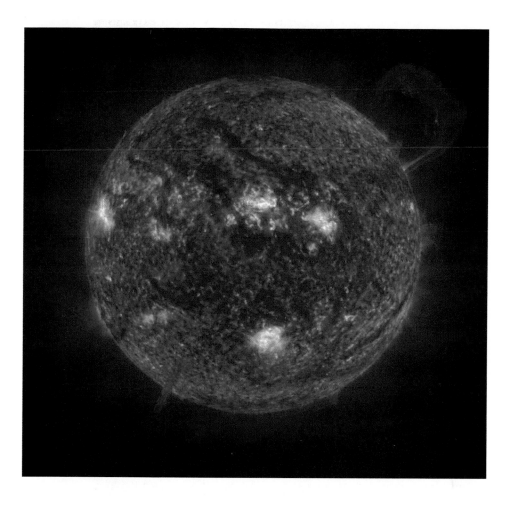

Extreme Ultraviolet Imaging Telescope image of a huge, handle-shaped prominence in the Sun's hot, thin corona. The hottest areas appear almost white, while the darker areas indicate cooler temperatures.

A traditional Nativity scene. *Adoration of the Magi* by Giotto di Bondone (1266–1336), Scrovegni Chapel, Padua, Italy.

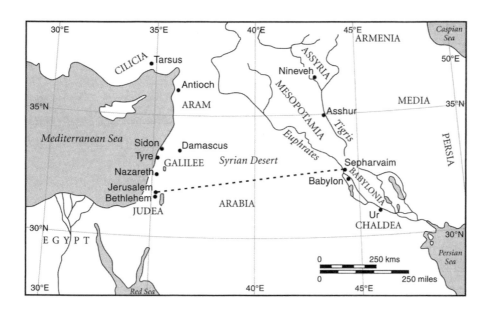

A map of the region traversed by the magi as they followed
the star of Bethlehem.

己　酉

已　酉

The Chinese pictograms *Chi-yu* (*top*) and *I-yu* (*bottom*)

Jupiter and three of the four Galilean satellites: Callisto, Ganymede, and Europa. At the time, Io was behind Jupiter as seen from Mars, and Jupiter's giant red spot had rotated out of view.

A turbulent region west of Jupiter's Great Red Spot.
The small, bright, white spots are believed to be thunderstorms.

Io in front of Jupiter

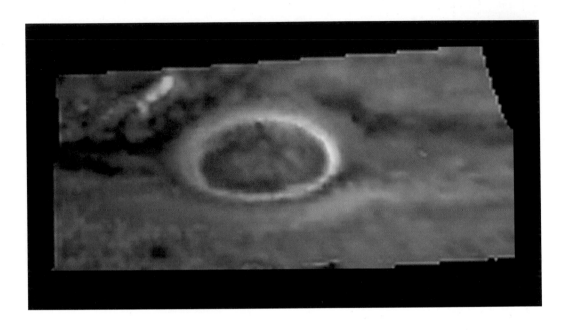

The first discrete ammonia ice cloud positively identified on Jupiter. Ammonia ice (light blue) is shown in clouds to the northwest (*upper left*) of the Great Red Spot (large red spot in middle of figure).

Layers of haze cover Saturn's moon Titan.

Saturn's moon Prometheus can be seen through the F ring.

Saturn displaying its familiar banded structure, with haze and clouds at various altitudes

Saturn's C ring (and to a lesser extent the B ring at top and left), revealing more than sixty bright and dark ringlets

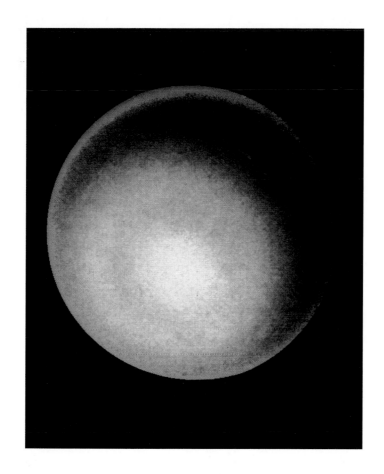

The atmosphere of Uranus

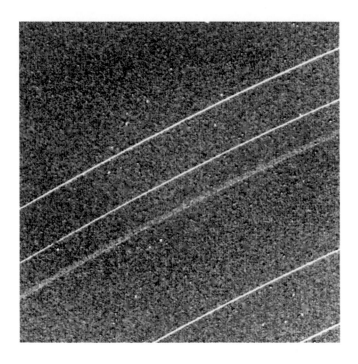

The Uranus ring system and lanes of fine dust particles

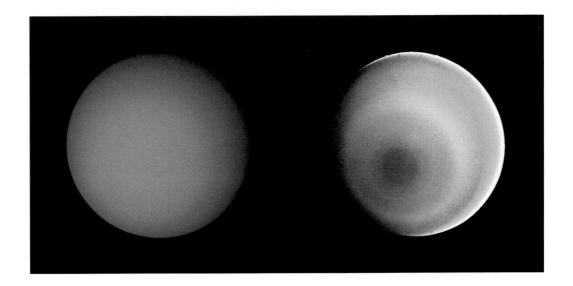

True and false color images of Uranus reveal a dark polar hood.

Buzz Aldrin as seen by Neil Armstrong, who is taking the picture

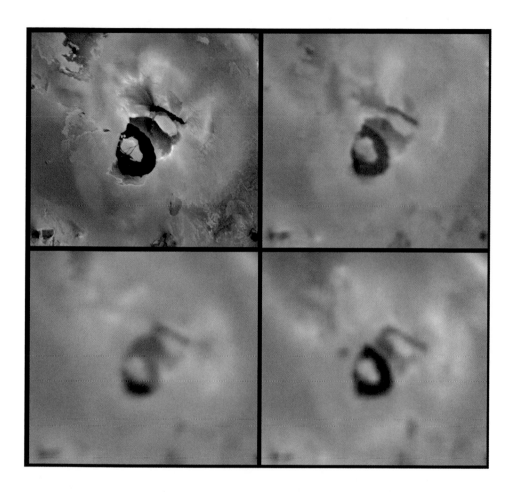

Changes near the Loki Patera, the central feature in Io's northern hemisphere. The large dark area might be a lake of liquid sulfur with a raft of solid sulfur inside.

Image of the Moon showing the distinct Tycho impact basin at the bottom of the image. The dark areas are lava rock–filled impact basins.

Crater Tsiolkovsky on the Moon

Closeup of sedimentary rock in Ganges Chasma, part of
the Valles Marineris trough system on Mars

The Valles Marineris of Mars

Syrtis Major (large, dark marking just left of center) can be seen
on the surface of Mars.

The first strong proof that meteorites could have come from Mars;
the black glass contains traces of Martian atmosphere gases.

Nobular nuggets that cover the Martian rock called "Pot of Gold." Data have shown that the Pot of Gold contains hematite, which can be formed with or without water.

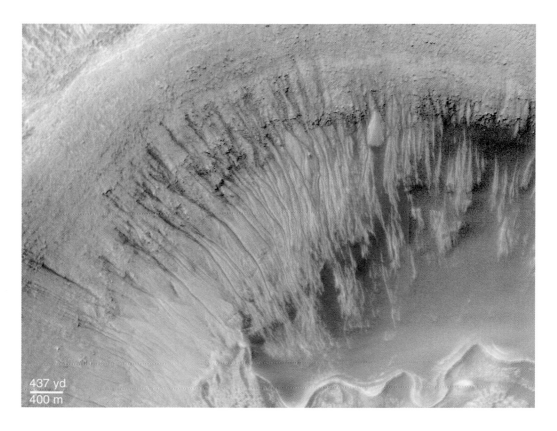

437 yd
400 m

An image of what has been called "weeping terrain" on Mars. The large streaks running down a Martian cliffside appear to be proof of the existence of recent flowing water on Mars, although a minority of skeptics believe that they may have been formed by liquid carbon dioxide under great pressure.

A representation of a possible underground Martian lake where subsurface ice is melted by the pressure of overlying rock and internal heat

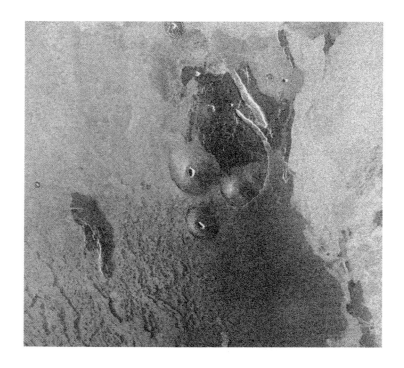

Some small volcanic domes on the flank of the volcano Maat on Venus.
The bright flows to the east are most likely rough lava flows, while
the darker flows to the west are probably smoother flows.

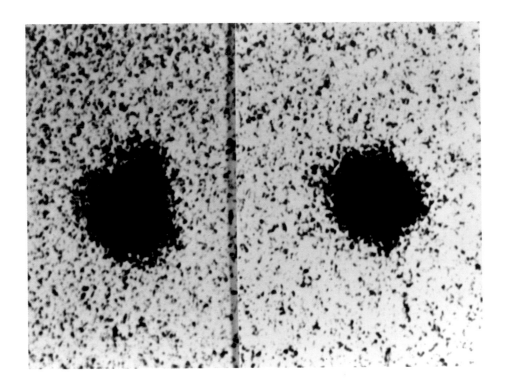

The discovery image of Charon. The bulge in the left image is Charon orbiting around the edge of Pluto; in the right image, Charon is invisible, having passed behind Pluto.

telescope is actually slightly fuzzy and cometlike. Or an astronomer might notice that the orbit of the object appears to be that of a comet and therefore check to see whether this is indeed what it is. Such situations initially posed a problem. Who discovered the comet? The astronomer—or telescope—who first saw and reported the object? Or the one who recognized it as a comet? It was ultimately decided that it should be the former. Should a specific question come up, the IAU's Small Bodies Naming Committee is the final arbiter.

Edmond Halley's Comet

Halley's comet has been seen in the sky for more than two thousand years. The Chinese were aware of it, but they gave it no special name. Indeed, over most of the centuries that Halley's comet has been observed, it has had no name, and its appearances have neither been expected nor welcomed. How, then, did the comet come to be associated with an astronomer who was only one of many people to observe it when it appeared in 1682? And how should one pronounce its name? Halley died over 250 years ago, so we cannot ask him. Three options have been proposed:

Is it Hay-lee, as in Bill Haley and the Comets?
Is it Hal-lee, as in the HAL9000 computer?
Or is it Haw-lee, rhyming with "jolly"—or, if you're English, with "poorly"?

Many people tended to use the first, probably by association. This option is most certainly wrong. Most of the Halleys listed in the telephone book, who were consulted in a survey in England in 1986, when the comet last returned, said that they pronounced their name the second way. But the pronunciation of British English has changed greatly over the past few centuries in comparison to American English (which would actually be far more familiar to the ears of Shakespeare).[3] How current Halleys pronounce their family name is therefore an unreliable guide. I, too, have always used the second pronunciation, and it appears that I also am wrong. According to the experts, the best historical evidence suggests that Halley himself pronounced his name the third way. It is this third pronunciation that is assumed in a poem written by an anonymous British astronomer at the time of the comet's 1986 appearance:

Of all the comets in the sky
The most famous is Comet Halley
We see it every seventy-six years
But this time rather poorly!

The 1682 Comet

The appearance of Halley's comet in 1682, the year that Halley himself observed it, seems not to have been very spectacular. The comet was first spotted by a group of Jesuit priests in Orléans on the morning of August 24, low in the dawn sky. It was already well inside the Earth's orbit, and, although brightening rapidly, it would soon become difficult to see in the morning light. Moreover, the comet was visible for less than a month—until only September 22. Halley, then twenty-six years old, first saw the comet on September 5 and then lost it on September 19. Three hundred years later we may wonder why the comet appeared for such a brief period. When it was seen for the last time the comet was fading as it got further from the Earth, but it was still close enough that it should have been visible to the naked eye. Such small mysteries turn up from time to time in the history of Halley's comet, and, as far as we can judge, it has not been equally bright on every visit. Sometimes the comet has been observed when it should have been far too faint or too distant to be followed; on other occasions, as in 1682, it disappeared, even though it should still have been bright enough and well enough positioned to be visible for longer.

Although the comet was not at its most brilliant in 1682, Halley was so impressed by the comet that over the course of his life he became increasingly fascinated with comets. In 1704, at the age of forty-eight, Halley accepted the chair of geometry at Oxford University and began a study that would have profound implications for our vision of the universe.[4] Halley examined the orbits of twenty-four comets that had been observed between 1337 and 1698, including the comet of 1682. He noticed immediately that the orbit of the comets of 1531 and 1607 looked remarkably similar to the one he had observed in 1682. He further noticed that the three comets were separated by 76-year periods, which led him to suggest that the three might be one and the same comet, returning at a standard interval. He therefore predicted that the same comet would be seen again in 1758. This was not, as is often stated, the first time that the return of a comet had been

What Our Forebears Knew

predicted. It was, however, the first time that firm mathematical evidence had been advanced to demonstrate that a comet would return.

Given that he would be 102 years old in 1758, Halley certainly had no illusions of ever seeing his calculations proved correct. In fact, Edmond Halley died in 1742, albeit at what even today would be the grand old age of eighty-six. Before his death, Halley, in grandiloquent style, made the following comment: "If the comet should return according to our prediction, about the year 1758, immortal posterity will not refuse to acknowledge that this was first discovered by an Englishman."

In fact, after Halley's death his calculations were checked by three French astronomers, Joseph-Jérôme de Lalande, Alexis Clairaut, and Madame Lepaute, and found to be slightly in error. Halley had slightly underestimated the comet's period, and the French trio revised the date of its next return—that is, its next perihelion, or closest approach to the Sun—to 1759. As the predicted moment of return approached, another French astronomer, Charles Messier, later known as the "ferret of comets," owing to his extraordinary success in discovering them, started to search along the path in the sky on which the comet should appear. Messier was hoping to discover what would have been his very first comet, but on this occasion he was disappointed. Although he located the comet on January 21, 1759, he then found to his immense chagrin that an Austrian astronomer named Johann Palitzsch had already reported the comet on December 25, 1758, without ever realizing that this was the long-awaited return predicted by Halley.

Messier was to take revenge for his defeat by dominating comet discoveries for the next forty years, despite stern competition from another Frenchman, Pierre Méchain. Officially, Charles Messier now has thirteen recognized discoveries to Méchain's eight, but when his independent discoveries of comets first seen by others are included the total becomes nineteen. Two stories illustrate both Messier's fanaticism and his rivalry with Méchain. On one occasion he was so wrapped up in staring at the sky as he walked along, hoping to spot a comet, that he fell down a well. On another, Messier had lost a comet to Méchain while attempting to cope with his wife's death. When a colleague offered him condolences, he replied, "Yes, alas, Méchain has robbed me of the comet." Only after a few seconds of contemplation did Messier realize that he had entirely missed the point of his colleague's sympathetic words.

The confirmation of Halley's prediction—the comet had duly appeared in

1758, even though it was not at its closest to the Sun until 1759—was to start a new tradition in astronomy. In 1760 the first comet to be named for the person who first discovered it (Comet Messier, not surprisingly) was added to the catalogues. But, appropriately, the first comet ever to be named was that of Halley. As with a small and select group of comets, all periodic, Comet Halley is named for the astronomer who investigated its orbit and showed it to be periodic, not for the astronomer who first saw it.

The 1835 Prediction Contest

We now know that Halley's comet was seen in 1531, 1607, 1682, and 1759—that is, four times over a period of 228 years. One might thus be tempted to believe that it would be a trivial exercise to work out when the comet appeared prior to 1531 and when it will return again. Divide 228 by 3 and we get 76. We should thus be able to work backward or forward at 76-year intervals for as long as we like. Right?

Absolutely wrong!

Let's try an experiment. We know that Halley's comet most recently appeared in 1986, so let's work forward from 1759 in 76-year periods and see when the comet *should* have returned most recently. We get predictions of 1835, 1911, and 1987. Not bad, but we are one year out. By the next return, which we know will take place in 2061, we will be two years out. In fact, much of the problem lies between 1835, when the comet *did* return on schedule, and 1910, when the comet returned after less than 75 years. This interval is not only the shortest here, but in fact it is the shortest interval between two returns of Halley's comet in over two thousand years.

Over the span of a few centuries the 76-year rule is a reasonably good guide as to when the comet will appear, but it fails to be absolutely accurate. One reason for this is the gravitational influence of Jupiter and Saturn, which alters the comet's orbit and thus increases or decreases the length of its orbital period. As a result, the comet can take anywhere between 74 and 79 years to orbit the Sun. The average over the last two thousand years has been just two days short of 77 years. Therefore, if you keep jumping back 76 years, the one-year differences will soon accumulate until the dates are way off.

As we have seen, prior to 1682, when Halley himself observed the comet, it

What Our Forebears Knew

was seen in 1607 and in 1531, and so one would expect it to have returned in 1455. There was no bright comet in 1455, but Halley did notice that one was observed in 1456, which he suspected was his comet, although he did not calculate its orbit. Actually, besides the one in 1456, there were two bright comets in 1457 and another in 1458, which somewhat confused the issue. It was not until 1783 that the French astronomer Augustin Pingré confirmed by calculating its orbit that the 1456 comet was indeed Comet Halley. The process of calculating backward from 1456 to determine when Comet Halley was seen in the more distant past would take astronomers two centuries of effort, inspiration, trial, and a lot of error. Even now, more than two hundred years after Pingré's calculations, we cannot say for certain when Comet Halley was seen prior to 240 B.C. But how about calculating forward? Once it was known exactly when the comet had appeared in 1456, 1531, 1607, 1682, and 1758, how accurately could its next return, in 1835, be predicted?

Time and time again astronomers have proved that, just like anyone else, they are capable of being extremely competitive. In 1758, the year Halley had predicted that the comet would next return, French astronomy was unrivalled in the world. When it was Johann Palitzsch who first spotted the comet, on December 25—nearly a month before Charles Messier found it—French astronomers were horrified that an Austrian amateur could snatch the first sighting of Halley's comet from them. As the 1835 return approached, a frenzied race accordingly ensued among astronomers in several countries to predict not the year but the exact day on which the comet would appear. Who would be closest? They realized that it was vital to include the gravitational force of the planets in their calculations. At a time when mathematical operations were done entirely by hand, on paper, working out the orbit of a comet could take weeks of effort. Each planet that was included enormously increased the complexity of the calculations, so coming up with a date for the comet was no trivial task. Even today, with modern computers and sophisticated computer programs, much depends on the skill of the orbit calculator, who must know which observations to accept and which to reject, and how to apply the software programs to the best effect.

The first to start was the French astronomer Marie-Charles Damoiseau, in 1820. Damoiseau painstakingly calculated the gravitational perturbations of Jupiter, Saturn, and Uranus on the comet and came up with an answer, 04:00 UT on November 17, 1835, which we now know was only eighteen hours too late. He subsequently realized, however, that because the comet had passed quite close to the Earth in April 1759 and would do so again in October 1835, it would be necessary to allow for the gravitational influence of the Earth as well. After another laborious round of calculations, Damoiseau arrived at a new result, which was more than twelve days earlier—and thus, unfortunately, less accurate than his initial estimate. A German astronomer, Jacob Lehmann, carried out similar calculations in 1835 and came up with a perihelion date at the end of November, more than ten days too late. Meanwhile, a French nobleman, the count of Pontécoulant, had made four predictions between 1830 and 1835, ultimately arriving at a time late in the evening of November 12—just three and a half days too early. Finally, another German, Otto Rosenberger, predicted midnight on November 12, a time almost identical to that finally proposed by the count of Pontécoulant. In the end, it was Pontécoulant who came closest, but, even though his calculations were as accurate as mathematics could make them, he was still wrong by an alarming three and a half days.

So all these predictions were quite close, but none was spot on, which must have been disappointing at the time. Of course, Damoiseau and his rivals worked under the rather severe handicap of not knowing that the orbit of Comet Halley was influenced by another giant planet, Neptune, as yet undiscovered. The fact that they were unable to take Neptune into consideration was especially critical in 1835. As Comet Halley receded from the Sun in 1759, Neptune was on the same side of the Sun as the comet, far closer than Uranus, which was on the opposite side of the Sun. In contrast, in the years before 1835 both planets were the same side of the Sun and therefore "pulling together." Under the circumstances, then, given an error of only three or four days we can now say that the calculations were brilliantly successful.

One might have expected that the discovery of Neptune in 1846 would resolve the problem. As in the years before 1835, prior to 1910 various astronomers attempted to predict the exact moment of the next perihelion, now with knowledge of the missing planet. The finest effort was made by two British astronomers, Philip Cowell and Andrew Crommelin. But despite their thoroughness and their careful checking and revision of their calculations their prediction was off by 2.7 days—not all that different from the error in 1835. Following the appearance of the comet, Cowell and Crommelin analyzed the discrepancy and figured out that at least two days in their error had to be due to reasons other than minor mistakes in calculation.

It would take astronomers more than sixty years to discover *why* these calculations were wrong. The problem was Comet Halley itself. Every time the comet is warmed by the Sun, jets of gas and dust appear at different points on its surface. These jets are not unlike the famous geyser Old Faithful, although they are less violent. Each jet acts like a small rocket, altering the comet's trajectory slightly, and it was these rocket forces that, in the end, frustrated all efforts to predict the precise movement of the comet. The same rocket forces make it impossible to calculate exactly where Comet Halley was in the distant past because they change over the years as old geysers die out and new ones appear. Even a few days' error in the position of the comet today may result in an error of several years over a period of two thousand years or more.

Pursuing Comet Halley over Three Millennia

Prior to the twentieth century astronomers had only two ways of identifying observations recorded in historical chronicles as referring to Halley's comet. The first was to calculate the orbit of a comet that had been observed in the past and see whether it appeared to be the same as that of Comet Halley. The second was to go back at approximately 76-year intervals and see whether a comet had been observed around the appropriate time that moved in the same way as a comet with Halley's orbit would. These approaches generated a number of successes. Orbits calculated for comets observed in 989 and 1301 were both close to that of Halley's comet and conformed to the approximate 76-year sequence quite well, so these were obviously both Halley. These two correct identifications simplified matters in that they extended the date of the last reliable observation of the comet back several centuries.

During the 1840s, two French astronomers made major contributions toward unraveling the history of the comet. Paul Laughier, who had already correctly identified the comet's appearance in 1301, showed that the comets observed in China in 451 and 760 and in the autumn of 1378 were also Comet Halley. Similarly, Édouard Biot demonstrated that the Chinese comets of 451, 989, and 1301 were Halley and added new Chinese sightings in 12 B.C. and in 684, 837, 912, 1066,

1145, and 1222 A.D. With this, every single appearance of Comet Halley back to 837 had been correctly identified, fourteen in all, as well as four previous sightings, in 760, 684, 451, and 12 B.C. Each success made it easier to add additional sightings. Just a few years later, a British astronomer, John Hind, tried to identify every apparition of Comet Halley from 11 B.C. to 1301 A.D. Although his success rate was quite high, six of his proposed appearances of the comet, including that in 11 B.C. and, more surprisingly, in 1222 A.D., were incorrect.

All these studies were published between 1842 and 1850. By then, however, astronomers had reached the limit of what trial and error could achieve. It would not be until 1907 that Cowell and Crommelin would carry out an amazing computational tour de force and thus take things a step further.

Cowell and Crommelin realized that the orbit of Comet Halley changed slowly with time. Since its appearance in 1066, which was recorded in the Bayeux tapestry, the period has decreased from 79.3 years (between 1066 and 1145) to a minimum of 74.4 years (between 1835 and 1910).[5] At the same time, its orientation has changed too. Technically, astronomers speak of the *longitude of perihelion*—the direction of the line between the comet's perihelion, or the point where it is closest to the Sun, and its aphelion, or furthest point—and the *longitude of the ascending node*—the point where the comet's orbit crosses the plane of Earth's orbit as the comet moves upward out of the plane (that is, as it passes from south to north of the ecliptic). Both these are measured as angles from the position of Earth in its orbit at the spring equinox, which serves as the Greenwich meridian of the solar system.

Both the longitude of perihelion and the longitude of the ascending node of Halley's comet have changed over the years, increasing by 24° and 28° respectively since 240 B.C., mainly as a result of the constant influence of Jupiter. What Cowell and Crommelin did was to measure the rate of change of these two values since 1531 and then laboriously calculate the orbit of the comet backward in time to 240 B.C., taking into account the perturbations by the six largest planets. Their calculations entailed an enormous number of mathematical steps, all done completely by hand. Indeed, their work is perhaps the largest and most complicated calculation ever carried out without a computer. Despite their prodigious efforts, though, and despite the fact that they were able to identify the comet observed in China in 240 B.C. as Comet Halley, the disheartening conclusion was that by 240 B.C. their calculations were almost a year and a half out.

What Our Forebears Knew

Modern Efforts

Since 1907, numerous attempts have been made to extrapolate the orbit of Comet Halley into the future as well as back into the past using ever more sophisticated computer programs. Two factors stymie all efforts, however. First are those tiny rocket forces, more correctly called nongravitational forces, which can be accurately determined only over the past four centuries and which have changed significantly even in that time. The second is that the comet repeatedly passes close to Earth over the millennia. An error of just a few days in the position of the comet substantially changes the distance at which the comet will pass by Earth and thus the extent to which Earth's gravitational field will alter the comet's orbit.

In 1977, Don Yeomans, a scientist at the National Aeronautics and Space Administration (NASA), set out to predict the exact time of Comet Halley's next return, in 1986.[6] By drawing on observations of the comet since 1607, and by including in his calculations a mathematical expression designed to allow for the influence of nongravitational forces, he was able, five years before the comet was spotted again, to estimate the exact time of its return with an error of just five hours. A five-hour error in 76 years is a phenomenal level of accuracy—to within 0.0008 percent, or one part in 130,000. When observations of the comet made in 1982 to 1984—after it was finally spotted once again—were added, Yeomans's error dropped to just half an hour. Predictions have already been made for the time of the Comet Halley's next two returns: July 28, 2061, at 17:05 UT, and then March 27, 2134, at 19:49 UT. Both predictions are likely to be accurate to within a few hours. We also know that the 2061 return of Halley's comet will again be visually disappointing. In 2134, however, the comet will pass very close to Earth about six weeks after its perihelion and will therefore be spectacular.[7] At the same time, because the comet will pass so close to our planet in 2134, its orbit will change, which leaves us unable to predict its subsequent returns with any degree of precision.

By combining modern software programs with ancient Chinese observations, we can also calculate the past orbits of Halley's comet with great precision, although the calculations cease to be as exact when we no longer have observations against which to correct the calculations. These new computer methods allowed the one missing return of Comet Halley, in 164 B.C., to be located. Because the comet's previous appearance, in 240 B.C., had been recorded by the Chinese, as had its subsequent one, in 87 B.C., it was relatively easy for software programs to predict where the comet would have been in 164 B.C. If the Chinese observed the comet in that year (as one would imagine they did), we have yet to find a record of their observations, but a reference to it has been found in a Babylonian tablet, translated by the British astrohistorian Richard Stephenson, which appears to indicate that the comet passed close to the Pleiades.

Modern software programs have also enabled us to demonstrate that the comet seen in 467 B.C. in China and in Greece, which had often been thought to be Comet Halley, most definitely was not. It is possible that a comet seen in 1059–58 B.C. was Halley's comet, but none of the other twenty-two comets observed in China and Greece before 240 B.C. can be identified with Halley, and it seems unlikely at this point that any other appearances of Comet Halley prior to 240 B.C. will come to our attention. Until Comet Halley returns in 2061, then, it will have been seen just thirty times throughout recorded history.

A Very Special Comet

Although Comet Halley is about ten times brighter than average, it is not the brightest of comets. It is, however, the only comet that returns at regular intervals roughly equivalent to a human lifetime and that is, at each return, visible to the naked eye. Alas, for us, it is no great consolation to learn that in 1986 the Halley's comet was the faintest it has ever been in over two millennia of observation. Only its appearance in 164 B.C. came close to being as unimpressive. Neither does it help to hear that had the comet returned just a few weeks later than it did in 1986, it would have been spectacularly bright, as it will be in 2134. In fact, in almost half (thirteen) of the thirty times known to us that Halley's comet has appeared in the sky it has reached negative magnitude, making it comparable to the very brightest of stars.[8]

When Comet Halley returned in 1986, it was greeted by a veritable fleet of space probes. Two Japanese probes, two Soviet ones, and the European probe Giotto, which became the star of the show, passed by it at different distances and gathered an immense amount of data. In addition, from 1982 to 1995, astronomers at observatories around the world tracked the comet, carrying out a prolonged series of observations. It thus seems safe to say that Halley's comet has been more extensively studied than any other comet in history. It was, of course, the first comet to have its history thoroughly investigated, as well as the first to have its composition analyzed. It was also the first comet to reveal its nucleus for study, which, perhaps above all, allowed us to confirm that the comet—and, by extension, comets generally—actually had a nucleus, something that had previously been no more than a theory that not all astronomers accepted. In short, we owe a great deal of what we know about these mysterious icy visitors from outer space to Comet Halley—a very distinguished comet that has rightly earned a special place in the annals of astronomy.

SUGGESTIONS FOR FURTHER READING

Popular Articles and Books

There is no shortage of material about Comet Halley that the nonspecialist reader will find both interesting and informative. In particular, the period around the time of the comet's recovery in 1982 and its appearance in 1986 produced an abundance of excellent articles. These can be found in back issues of popular magazines such as *Sky and Telescope*.

Peter Lancaster-Brown, *Comets, Meteorites and Men* (London: Robert Hale, 1973).

> *A wonderful book that, although written before the 1986 return of Halley's comet, is full of fascinating historical information about both Comet Halley and comets in general. Still available through Internet bookstores.*

More Advanced Reading

Donald K. Yeomans, *Comets: A Chronological History of Observation, Science, Myth and Folklore*, Wiley Science Editions (New York: John Wiley and Sons, 1991).

> *A dense and comprehensive study of how our knowledge of comets has changed over the ages. A lot of detailed science, particularly in relation to Comet Halley.*

R. Reinhard and B. Battrick, *Space Missions to Halley's Comet*, ESA-SP 1066 (Noordwijk, The Netherlands: ESA Publications Division, 1986).

> *A book specially prepared by the European Space Agency for the encounters of the Giotto space probe with Halley's comet in March 1986. The volume contains one excellent historical article as well as much useful material about the aims of the space probes generally and, in particular, the Giotto mission.*

On the Internet

The Halley Multicolour Camera
www.linmpi.mpg.de/english/projekte/giotto/hmc/

> *This is the official page for the camera of the Giotto probe that visited Comet Halley in March 1986. Spectacular images, links, and information concerning the probe's findings all in a compact page make this a valuable starting point for readers who would like to learn more about the comet without being overwhelmed by detail.*

Comet Halley
www.nineplanets.org/halley.html

> *A page with history, links, and images from the Lunar and Planetary Laboratory's Nine Planets Web site.*

Catalog of the Scientific Community: Halley, Edmond
http://es.rice.edu/ES/humsoc/Galileo/Catalog/Files/halley.html

> *A detailed biography of Edmond Halley that contains many little-known details about his life and work.*

IP/Halley
http://cometography.com/pcomets/001p.html

> *The American Meteor Society page for Comet Halley. Includes a brief history and some excellent images of the comet's 1986 return.*

PART 2

Looking around Our Solar System

THE SPACE AGE BEGAN IN SPECTACULAR FASHION WITH the launch of Sputnik I in 1957. At the outset progress was almost unbelievably rapid. It was less than twelve years from the time Sputnik I orbited the Earth to the moment when Neil Armstrong left the first human footprint on the Moon. By the late 1960s enterprising travel agents were already selling tickets for moon flights at what would now be bargain prices, and there were grand plans to put an astronaut on Mars by 1980. Such was our enthusiasm that when Stanley Kubrick's famous film *2001: A Space Odyssey* appeared in 1968, around the time that Apollo VIII was launched, many people believed that its scenes of the exploration of the solar system could well be reality by 2001. Kubrick even felt obliged to make inquiries about insurance to protect himself in the event that Apollo VIII discovered evidence of alien visitors to the Moon much in the way the film portrays—

thereby preempting his plot, along with a sizeable share of his audience.

But instead of continuing to step forward into space, in the decades that followed we retreated. In February 2003 the Columbia tragedy—the fourth fatal accident to occur in the American and Russian space programs—issued a powerful reminder that spaceflight continues to be complex and dangerous. Inevitably, there are those who ask whether it is all worth it. Far from conquering the solar system, we face the very real danger that we could make the catastrophic decision to withdraw from space completely.

But the solar system is our backyard. Over roughly the past half century unmanned probes have investigated, in greater or lesser detail, the surface of the Moon, of Venus, of Mars, and of Titan, as well as the atmospheres of the outer planets. Space probes have also visited comets and asteroids, and, of course, the Moon has been explored by astronauts. Moreover, as we will see, in recent years astronomers have developed many ingenious techniques for learning about places in the solar system that we may never be able to visit. We are fortunate to have such techniques, for our neighbors in space have much to teach us. In particular, Mars may hold the key to what is arguably the most compelling of all the many mysteries in the solar system: the origin of life itself.

Quite apart from our desire to resolve purely abstract questions, however, reaching our nearest neighbors in space is equally a matter of pragmatism. To protect the Earth's natural resources from utter depletion, and ourselves from the threat of extinction, we need to explore the possibilities offered by the other planets. It is clearly in our best interests to discover how to make use of the resources they could provide. In fact, to choose not to do so may well be to court disaster.

CHAPTER 5

Our Moon: Nearest Neighbor—and Hot Property?

The most spectacular object in the night sky is one we take for granted. Many astronomers complain about the Moon because its light overwhelms the faint, distant objects at the edge of the universe that they want to view. Most people don't give the Moon a second glance, though—except perhaps at moonrise, when the huge orb is such a majestic sight on the horizon that even the hardest hearted will have their eyes drawn to it.

On a July evening now over thirty-five years ago—to be precise, on July 20, 1969—the eyes of the whole world were on the Moon. Two lonely explorers in a flimsy craft, so delicate they could have punched a hole in its walls, landed gently on the Moon's surface. In England, inspired by the BBC's attempt to make the flare of the lunar module's rocket motor visible to viewers by attaching a TV camera to a 12-inch telescope, a nine-year-old boy took his binoculars out into the garden to look at the fat crescent Moon and share in that moment. Neither I nor the BBC saw the rocket's flare as Apollo XI descended toward the lunar surface, but this detail was quickly forgotten because we were all sharing history. Then, on July 21 at 2:53 P.M., London time, Neil Armstrong climbed clumsily down the lunar module's ladder in his bulky space suit, touched the Moon's surface with his boot, and said: "That's one small step for man. One giant leap for mankind."[1]

During the 1960s, the space program built up momentum as the United States attempted to beat the Soviets to the Moon, to the point that NASA became one of the country's biggest employers. Nobody anticipated that spaceflight and even trips to the Moon would become so routine and boring to the general public that, by the time the Apollo XVII mission took place, in 1972, viewers were telephoning the networks to complain that reruns of *I Love Lucy* had been canceled so that the latest moonwalk could be broadcast. Politicians sensed the public mood: space was no longer exciting.[2] Thus, rather than landing on Mars by 1980, as NASA had originally envisaged, in addition to establishing a permanent moon base, we retreated from space. In the years that followed, NASA suffered massive cuts in budget and personnel. By the 1990s, NASA officials had to admit that if they were told to get back to the Moon within ten years, they would be incapable of doing so. Whatever effort might be made, and whatever budget allocated for that effort, it would not be possible to repeat what was done from scratch in eight years in the 1960s. Moreover, although the Apollo landings did help us to solve some of the riddles about the origin and nature of the Moon, in some ways they made the Moon even more mysterious. Those hundreds of kilograms of moon rock posed more questions for scientists than they answered, at least for a time. Only in the past few years have we finally begun to unlock some of the Moon's secrets.

Our Puzzling Moon

A cursory glance at our solar system is enough to reveal that our Moon is a somewhat anomalous body. As of February 18, 2005, there were 139 known natural satellites of the planets (as opposed to orbiting space probes), although their number has been increasing rapidly as yet more small satellites of the outer planets are discovered. Our Moon, with a diameter of 3,476 kilometers, is the fifth largest of these satellites. Three of the Galilean moons of Jupiter, the largest planet in our solar system, are bigger than the Moon, as is one of Saturn's satellites. Of the four inner planets, Mercury, Venus, Earth, and Mars, only Earth has a satellite of any serious size. Mercury and Venus have none, while Mars has two tiny moons that appear to be captured asteroids. None of the moons of Uranus comes close to the size of our Moon, and even the largest satellite of Neptune, Triton, is significantly smaller.

Moreover, Earth's Moon is considerably larger—five times more massive—than the planet Pluto. It is also much closer in size to its parent planet than any other natural satellite in the solar system, leaving aside Pluto and its relatively outsized moon, Charon. The next largest satellite in relation to its planet is Saturn's moon Titan, but Saturn is 4,220 times more massive than Titan, whereas Earth's mass is only 81 times that of its Moon. Most satellites are less than one ten-thousandth of the mass of their parent planet, and the majority are a millionth or smaller. At least in this sense, the Earth and the Moon seem more like a double planet than a planet and its satellite.

The Moon also has another peculiarity that is unique within the solar system. Although we say that the Moon orbits the Earth, the situation is not quite that simple. If we draw a diagram of the Moon's orbit relative to the Earth's, we discover that owing to the strength of the Sun's gravitational field, the Moon is constantly trying to fall into the Sun rather than the Earth, although it has enough lateral velocity that it never actually does so. In other words, the Moon is going around the Sun more than around the Earth, although it is in a permanent orbital dance with the Earth at the same time. The Moon's anomalous orbit results from its distance from the Earth, which is far greater than is the case for most planetary satellites. The mean distance of the Moon from Earth is sixty times the Earth's radius. This is more than double the distance between Jupiter and its outermost moon, Callisto, almost triple that between Uranus and its outermost moon, Oberon, and four times greater than the distance from Neptune to its large moon, Triton. In fact, of the solar system's major satellites, only Iapetus, the outermost of Saturn's major satellites, is as far away in relation to its parent planet, and Iapetus is exceptional in being more than twice as far from Saturn as any other of the planet's major satellites. However, while the Moon is a long way from us now, in the past it was considerably closer. The Moon is spiraling rapidly away from Earth, at a rate of about 3 centimeters a year. This may not sound like much, but over a million years it amounts to 30 kilometers. In the five billion years since our planet formed, the Moon would have moved 150,000 kilometers further away from the Earth if the rate had remained constant, although it has not. It was faster still when the Moon was closer.

But how did the Moon get to where it is in the first place?

The Moon's Mysterious Origin

Scientists assume that the larger satellites of the planets were formed at the same time as their parent planet as a miniature solar system. By contrast, the smaller satellites are almost certainly captured asteroids that may someday escape again. In the case of Jupiter, for example, many of the small satellites are so very tenuously held in place by the force of gravity that they are only temporary denizens of the Jovian system. The orbits of its outer moons are not even approximately elliptical because the Sun's gravitational pull struggles with Jupiter's own for dominance. In this fight, the Sun will always be the final victor. The Moon presents a problem for this theory, however, for it is clearly unlike the Earth in many ways. It is much less dense, for example, which implies that it does not have a large iron core. This is hard to explain if the Moon and the Earth formed out of the same cloud of material. We will return to this problem later, but let us first look at the theories that have been devised to explain the origin of the Moon.

For many years the most popular theory was that proposed by the great naturalist Charles Darwin. Darwin realized that the Moon's relatively low density posed a serious problem. The density of the Earth is 5.57 times that of water, whereas the Moon is only 3.34 times as dense. Granted, the Earth's huge mass compresses its core considerably and thus greatly raises its density, an effect that does not occur in the Moon. All the same, it is clear that the Moon cannot be made of the same material overall as its parent planet, because the difference in density is far larger than could be accounted for by the compression effect. But Darwin also noticed that the density of the Moon is similar to that of the outer layers of the Earth, the lighter material that floats on the denser material within. Darwin therefore suggested that the Moon and the Earth were originally a single body. This body, he surmised, rotated rapidly, so rapidly that the Moon broke away from the Earth's outer layers like a droplet thrown off a rapidly rotating ball. The Moon left a scar behind when it broke off, in the form of the Pacific Ocean.

Darwin's theory was accepted for many years. After all, it explained many things about both the Earth and the Moon. Had the Moon been formed from the Earth's light outer layers, this would account for its comparatively low density and its lack of any significant quantity of iron. Darwin's theory also explained the origin of the Pacific Ocean, the largest and deepest ocean on our planet, and why the Moon is so large in relation to the Earth. In order to stabilize the system, the droplet thrown off by such an unstable and rapidly rotating planet as Darwin envisaged would obviously have to be a big one.

Other versions of this theory were subsequently proposed. One of these suggested that sympathetic vibrations between the tidal budge in the Earth, which is caused by the Sun's gravitational pull and which moves along as the Earth rotates, and the Earth's natural frequency of vibration could have produced a catastrophic resonance. This may sound complicated, but examples of this phenomenon can easily be found. The Millennium Bridge over the Thames in London, for example, had to be closed almost immediately after it opened because of resonance effects that set it swaying uncontrollably in the wind. Worse still, the Tacoma Narrows Bridge in the United States—nicknamed "Galloping Gertie"—ultimately collapsed because of this problem. Soldiers therefore break step when crossing a bridge to avoid making it vibrate and possibly fall. An opera singer shattering a glass with a pure sustained note is another example. These are all resonance phenomena.

Unfortunately, Darwin's theory has all manner of problems. For one thing, the Pacific Ocean is far smaller in volume than the Moon (and, of course, we now know that the Pacific Ocean was formed by continental drift). For another, had the Moon been flung off a rapidly spinning Earth, its orbit would necessarily be in the plane of the Earth's equator, but it is not: it is inclined by some 5 degrees. Moreover, if the Earth were spinning fast enough to cause a substantial part of its mass to be flung off, the Earth itself would probably disintegrate. The killer, though, is that the angular momentum of the Earth-Moon system is wrong: the Moon is far too massive to have been contained in even a very rapidly spinning primitive Earth. Whereas ordinary momentum refers to objects moving in a straight line, angular momentum is the amount of spin that a spinning object possesses, whether the object is turning on its own axis or in orbit around another object. Ordinary momentum is simply the speed of an object multiplied by its mass ($M = s \times m$). Angular momentum, however, is the product of the object's speed times its mass times the distance ($AM = s \times m \times d$)—with distance referring to the separation between the object and whatever it is spinning around. What is critical here is the principle of the *conservation of angular momentum*. This principle states that, although speed and/or mass and/or distance can vary, the total amount of angular momentum cannot change. Thus, assuming the mass of the object remains the same (when would it not?), then if the distance decreases, the speed of the spin must increase, and vice versa.

As it is, our Moon has a specific mass, stays roughly the same distance from the Earth, and orbits the Earth at more or less a fixed speed. But, in order to test Darwin's theory, supposing we were to draw the Moon back in and reattach it to the Earth. As the distance decreased, the speed of the spin—in other words, the speed at which it orbits the Earth—would necessarily increase, so that the overall amount of angular momentum could remain constant. The problem is, however, that by the time the Moon—which is not only a large object but a long distance from us—had been pulled all the way back to the Earth, it would necessarily be spinning around the Earth very fast, so fast that if it were rejoined to the Earth and its spin thus combined with that of the Earth, the Earth would break apart. You cannot pack so much angular momentum—that is, spin—into a body as small as the Earth without catastrophic results.[3]

So, if the Moon did not break off from the Earth, could it have formed at the same time as the Earth and in the same part of space? As we have seen, a similarity exists between the Moon's material composition and the composition of the outer layers of the Earth. Here the Apollo missions provided a key clue, having to do with the isotopes of oxygen. Oxygen is found in three forms: with a mass of 16 (the most common), 17, and 18. Oxygen 17, or ^{17}O, has one extra neutron, and ^{18}O has two, but chemically they are all identical. The relative amounts of these isotopes differ in specific bodies within the solar system. If the ratios of oxygen isotopes are closely similar in two bodies, then they probably formed together or at least in the same region of space. When scientists tested samples of moon rock gathered by the Apollo astronauts, they discovered that the ratio of the oxygen isotopes is the same on the Moon and on Earth. This strongly suggests that the two formed from the same material, which lends support to what is known as the theory of coaccretion. According to this theory, the Moon and the Earth were formed at the same time and at the same point in space from a common cloud of dust and gas. But instead of producing just one large mass, as would have been the case for the other planets, this cloud of dust and gas developed into two objects that never joined together.

The idea is attractive because not only does it explain the Moon's large mass, but it also accounts for why the Moon and the Earth are so alike in composition. Unfortunately, there are two problems this theory cannot satisfactorily explain. First, if the Moon and the Earth were formed of the same material, how is it that the Moon was left without, or almost without, an iron core? If both were created from the same cloud of material, both should have had a roughly equivalent quantity of iron. It is difficult to imagine how the other elements were evenly distributed, but iron was not. The second problem, and again the big killer, is angular momentum. As a large body at a great distance from Earth, the Moon has a huge amount of angular momentum—basically, the amount of its spin multiplied by its mass. Getting that much angular momentum out of a cloud of dust is no easy matter.

The third of the mainstream theories is that the Moon formed elsewhere in the solar system and was later captured by the planet Earth. We know that after the solar system took shape—with the planets forming from the accretion of a huge cloud of dust and gas into individual chunks of material that increased in size by colliding with each other and combining—a large amount of loose material, called planetesimals, remained. Some of this material was in the form of large asteroids, and part was in the form of bodies as large as planets but not in stable orbits. We know that Jupiter and Saturn, and probably Mars, Uranus, and Neptune, captured some of these bodies, which then became their smaller satellites.

In general, the planets are less dense as we move away from the Sun. Mercury is apparently more than two-thirds iron. Venus, although less massive than Earth and therefore less compressed at its center, is almost as dense as Earth and must also have a large amount of iron in its core. Mars is far less dense—only a little more so than the Moon. The asteroids have a density similar to that of the Moon, which is to say that they are much less dense than the inner planets, but we know that they still contain a reasonable quantity of iron from the fact that many meteorites—fragments of asteroids that have split off—are made of iron. Jupiter has a much lower density still, while Saturn has such a low density that it would float in a bowl of water (if we could find one big enough). Given this progression, it seems reasonable to conclude that the Moon may have formed somewhere in the asteroid belt.

This theory avoids some of the problems of the other two, but it creates others. The biggest problem is still the Moon's apparent lack of an iron core. It is hard to imagine that a body as large as the Moon could form with little or no iron at its core, especially in the asteroid belt, given that many asteroids are composed partly of iron. A second problem is that we know that the interior of the Moon was molten in the past and, as far as we can ascertain from the data on moonquakes gathered by the instruments that the Apollo astronauts left on the Moon, appears still to be at least partially molten. Where did all this heat come from if the Moon, unlike the Earth, never had a large amount of heavy radioactive material in its interior—dense elements such as uranium and thorium that would sink down to the Moon's core, melting it with the heat of their radioactivity? In the great wave of collisions that took place about 3,500 million years ago, the great lunar seas such as the Mare Imbrium were formed, and these craters were flooded with lava flowing out from the interior of the Moon. The existence of lava implies that more than a thousand million years after it formed the Moon was still hot enough inside that it consisted of only a thin crust of hard material on top of a completely molten interior. Being such a small body, however, and one lacking a significant source of interior radioactive heat, it should have cooled quickly.

Yet another problem is plausibility. If the Moon had come in from, say, the asteroid belt, it would not be easy for Earth to capture it, given that it would be a massive object that would pass by the Earth at a velocity of several kilometers per second. The process would require that the asteroid destined to become the Moon make many thousands of passes close to the Earth, each of which would modify its orbit slightly, until it was finally brought into orbit around the Earth. Even then, as in the case of Jupiter's outer satellites, the Moon's orbit would probably be highly eccentric. All the moons in the solar system that were presumably captured by their parent planets are actually quite small bodies, and many are in such unstable orbits that they may escape again relatively quickly. It is thus unlikely that the Earth could capture a body as large as the Moon and even less likely that this could happen quickly. In fact, the process would take hundreds of millions of years.

In other words, all the established theories of how the Moon formed have serious and probably fatal problems.

A Giant Leap for Mankind or a Fool's Errand?

The year 2002 marked the thirtieth anniversary of the final Apollo moon landing. On December 15, 1972, Eugene Cernan became the last astronaut to leave the surface of the Moon. What started as the greatest human adventure of the twentieth century ended on a depressingly low note. At one time it seemed that the Apollo program would dominate the 1970s as the most exciting scientific frontier in history—taking us from that one small step into a systematic exploration

Looking around Our Solar System

of a new world. In the end, the events of the 1970s dominated the Apollo program. How many people thirty years old or younger can even name the first astronaut to step on the Moon, let alone the last? In fact, in 2001 a Fox News program reported that a significant minority of Americans—around 20 percent, according to the program—believe that the Apollo moon landings never took place. Rather, they were actually filmed on a desolate stretch of land near Houston (home of Mission Control and NASA's Johnson Spaceflight Center) with the help of Hollywood special effects.[4] This imaginative theory notwithstanding, the fact remains that, to many, the space program now appears to have been little more than an obscenely expensive frivolity of the crazy sixties. The Apollo program cost $24 billion—that's nine zeros on the end. Why did we do it?

The Apollo program was motivated by a number of factors. Political pride was an important one. In the years following it became popular to deny that the race to the Moon had ever existed. Finding themselves on the losing side, the Soviet authorities did little to discourage this denial, and few people in the United States and Europe were ever aware how close NASA came to losing the race to the Moon. Had Apollo VIII failed for any reason, a Soviet astronaut might very well have been first to step on the Moon. On December 6, 1968, two weeks before Apollo VIII was launched, cosmonaut Pavel Belyayev arrived at Baikonur ready to play his part in what we now assume was a plan to orbit the Moon starting on December 8. No launch was attempted, however, even though the Soviets were under intense pressure to beat the Americans. Then, shortly after Apollo VIII's flight, the Soviets witnessed their own booster—the rocket that would have lifted the manned space capsule into orbit, on the first leg of its journey to the Moon—blow up on the launching pad. This untimely setback marked the end of the Soviets' program for manned spaceflight to the Moon.[5]

The romantic view is to imagine that we traveled to the Moon in order to learn. Science was indeed a motive, of course: scientists were sure that studying the Moon's surface would answer many questions not only about the Moon but, indirectly, about the formation of the Earth as well. It is salutary to remember, though, that the serious science missions did not get underway until the Apollo XV flight and that the one and only scientist to go to the Moon, geologist Jack Schmidt, did not fly until the final Apollo mission, Apollo XVII.

The six Apollo moon landings brought back 2,196 samples of dust and rock, weighing a total of 381.7 kilograms. The last three science missions contributed

75 percent of this amount. In other words, each of the last three missions returned about as much moon rock as the first three flights put together. Particularly because the astronauts who participated in the first moon landings had no formal training in geology, had the landings ended with Apollo XIV the scientific results would have been poor. Nonetheless, those final three missions did produce some striking discoveries that yielded fundamental information about the history, structure, and formation of the Moon. Once the flag waving was out of the way, the Apollo astronauts were able to carry out an impressive amount of serious research on the Moon. Even though at this point the most spectacular science has been wrung out of both the moon rocks and the scientific instruments, such as seismometers, that were left on the Moon by each mission, a large fraction of the rocks remain to be studied and will provide work for geologists for years to come. A few special samples have not yet even been opened because scientists are waiting for the appropriate new techniques to be developed that will allow them to be opened without damaging the information they contain.

A third motive was undoubtedly economic. At the peak of the space program, some four hundred thousand people worked for NASA, and millions more depended on the space program for their livelihood. The initial impetus for the space program had largely to do with the cold war. President Kennedy had asked his scientific advisors to suggest something—*anything*—at which the United States could beat the Soviets, and, of the four projects they had proposed (which included a manned space station), he had selected a lunar landing. At least for the politicians, then, if not for the scientists, the spirit of scientific discovery was a tiny factor at best. The question of motivations aside, by the late 1960s the Apollo program had become a huge factor in the American economy. Once in place, it generated its own momentum.

But what, finally, did we get out of the Apollo program—and was it worth it? We originally went to the Moon primarily to prevent our adversaries from gaining a critical foothold in space. If our original motivation was political, however, once it became obvious that the momentum in the space race had shifted definitively toward the United States, other motivations came to the fore. We continued to fund the space program partly to seek answers to scientific questions and partly because, like climbing Everest or running a marathon or sailing around the world single-handed, it was a challenge. But we were also spurred on by a sense of wonder. Even the most hard-bitten politicians were stunned, first by the images of the far side of the Moon and then by the sight of men standing on its surface. For the general public, the Apollo landings lent an air of triumphant adventure, as mission after mission attempted something that had never previously been accomplished. Moreover, thanks to television, the excitement of this adventure was brought into the homes of millions of people around the world. The family of a young school friend of mine (we were both ten years old at the time) was one of the first in my village to have a color television, and I was regularly invited to watch the missions on it, the color images heightening my amazement at what I was seeing. Truly, the space program was the first great voyage of discovery in which the entire population had the opportunity to participate.

One could argue that if all those billions of dollars had been invested at home, then poverty would have been licked—although a more cynical view is that very few of those dollars would have found their way into humanitarian programs, while the rest would have gone to fighting the cold war or to the conflict in Vietnam. Others have claimed that the only thing we got out of the Apollo program was nonstick frying pans. But this is unfair. NASA documents over six hundred technological applications of the research that went into the space program (teflon included). That said, the value of technological applications, or "slop over," can be a matter of debate. Taken on their own, do these six hundred new items of technology justify the expense of going to the Moon? Probably not. Given the cost of the missions, even the advances in medical technology associated with the Apollo program might seem like a relatively trivial payoff. At the same time, one might put the question to the hundreds of thousands of people whose lives have been saved by the sophisticated medical instrumentation of intensive care units, most of which was originally developed to monitor the health of astronauts who were four hundred thousand kilometers from the nearest doctor.

The skepticism of some notwithstanding, above and beyond anything else the dollars spent on the Apollo program bought us raw knowledge and new technology, and that is beyond price. Thanks to the Apollo program, a large team of scientists and engineers were obliged to explore ways of doing things that had never been tried before. Their work led to advancements in electronics, computer technology, materials science, cryogenics, medicine, and bioscience, just to name a few.[6] If one adds in the benefits of the Apollo program to technology as a whole, the payback becomes incalculable. Spending money to learn how to do things is very rarely a wasted effort. In fact, investment in technology almost always pays off quite rapidly. It is instructive, for example, to look at the correlation between a country's degree of development and the fraction of its budget that it dedicates to pure research.

The Post-Apollo Moon

It goes without saying that the Apollo missions fundamentally changed our knowledge of the Moon. Although the amount we know about the lunar interior is still sadly limited, the moon landings have provided invaluable information about the Moon's material composition. The first landings, which took place in the flat, dark regions known by the Latin term *maria,* the plural of *mare,* "sea," confirmed our expectations that these areas genuinely are seas—but not of water. It came as no real surprise that they are seas of frozen lava. The lava is composed of a kind of igneous rock known as basalt that formed when magma welling up from the lunar interior cooled. We know the magma cooled rapidly because the crystals in it are tiny; had it cooled slowly it would have had time to form much larger crystals. We also know that the rocks are extraordinarily desiccated. On Earth, vulcanism—volcanic activity—releases large quantities of water vapor, which alters the chemistry of the rock. Not a trace of water or of its action exists, however, in any of the rocks brought back from the Moon.

In addition to volcanic activity, there have also been moonquakes, which appear to occur at two depths. Some of these are shallow quakes, close to the lunar surface. These have at times been caused by the impact of meteorites, two of which, in July 1972 and in May 1975, appear to have weighed as much as 5 tonnes. Others were brought about by rockslides or by thermal stresses breaking the surface rocks. But most of the quakes have occurred deep inside the Moon at a depth of about 1,000 kilometers. Overall, the amount of seismic activity is low, and none of these moonquakes would have been noticed by an astronaut standing on the Moon's surface. The deeper quakes seem to be due to tidal stresses produced by the Earth's gravitational field, which distorts the Moon's shape. Interestingly, the depth at which they occur appears to mark the boundary of an inner, partially molten region of the Moon called the asthenosphere. The discovery of this was rather unexpected, for the Moon was thought to be completely cold throughout.

Perhaps the most surprising result of the Apollo landings, however, was the discovery, initially made by the Apollo XV mission, that the amount of heat flow-

Looking around Our Solar System

ing out of the lunar interior is far greater than had ever been anticipated. This finding was regarded as so surprising and important that the experiment was to have been repeated by Apollo XVI. This never happened, because a cable connected with the experiment broke during deployment, when one of the astronauts tripped over it, and despite considerable efforts could not be mended. But the Apollo XVII astronauts successfully confirmed the results. As the experiment further demonstrated, the heat does not come from the core of the Moon but rather from a quantity of radioactive elements in its outermost 300 kilometers.

The Apollo landings also revealed that the Moon has no global magnetic field—a compass would be useless there. Some rocks do seem to have been exposed to a fair degree of magnetism in the past, however, that was apparently at its strongest about a billion years after the Moon formed. The discovery of this magnetism suggests that there might once have been a molten iron core, but, if it existed, it must have been small and cannot now be detected. Instead, the Moon consists of a thick mantle of light rock with a partially molten core, on top of which lies a crust that is around 70 kilometers thick. The outer part of the crust is a layer of soil, called the regolith, up to 15 meters deep, which is made of crushed rock from the tremendous lunar bombardment of the past. The rocks are similar in composition to those of the outer layers of the Earth, but the Moon does not seem to have formed either with the Earth or from the Earth. Thus, even with all we have learned from the Apollo landings, the origin of the Moon remained a mystery.

Above all, perhaps, what the Apollo missions showed us is that only one thing is missing from the Moon that would allow it to support a self-sustaining colony, and that is hydrogen. The Moon provides everything else that is required, including silicon for generating solar power and titanium and aluminum for construction. In addition, huge amounts of oxygen are bound up in the rock. At the same time, the Apollo landings also revealed that the Moon is totally sterile, apart from some bacteria that had survived inside the camera of the Surveyor 3 spacecraft, one of the unmanned probes that had soft-landed on the Moon, in 1967, as part of the preparations for the Apollo missions. (It landed on April 20, and between then and May 3, 1967, when it ceased operation, made 6,326 photographs of the lunar surface.) The Apollo XII astronauts took the opportunity to visit the Surveyor 3 probe, as they landed only 300 meters away from its resting place in the Ocean of Storms.

Returning to the Moon

In the late 1960s NASA had plans for a moon base in the 1970s and a manned flight to Mars as early as 1981. These ambitious plans were rapidly dismantled, however. From 1972 to 1994 the United States did not send even an unmanned mission to the Moon. The retreat from the Moon has been so total that, despite President Bush's announcement early in 2004 of plans for a future manned return to the Moon, it now seems highly optimistic to expect astronauts to set foot there before the fiftieth anniversary of Apollo XI, in 2019.[7] Nonetheless, the past few years have seen at least a timid return to the Moon. The Clementine probe, launched in 1994, and then, in 1998, the Lunar Prospector probe have carried out geological mapping programs and surveys of the Moon's resources and have returned some unexpected results. These probes were intended in part to prepare the way for active use of the Moon in the future, and we will learn more about them in chapter 11.

Of course, one of the key questions that scientists wished to settle was that of the origin of the Moon itself. As far as the three standard theories were concerned, however, the results of the Apollo missions were initially very perplexing. The moon rocks proved to be similar in composition to the Earth's outer layers, and yet it had already been demonstrated that the two theories capable of explaining such a similarity—those of fission and of coaccretion—were fundamentally flawed and hence untenable. But if the capture theory were correct, then the similarity should not exist. Then, in 1975 and 1976, two teams of scientists who were puzzling over the Apollo results independently made a daring suggestion about the Moon's origin. Perhaps the Moon had been part of the Earth after all. Many of the difficulties encountered by earlier theories concerning the origin of the Moon could be explained, they argued, if the Moon had been formed in a catastrophic collision between the Earth and a huge asteroid, one roughly the size of Mars. The original planet Earth would be totally destroyed by the impact, which, it was estimated, would have occurred 4.6 billion years ago, shortly after the original Earth was formed. If the impact had been off center, part of the outer layers of the Earth would be hurled into space, while the Earth's core and that of the asteroid would have combined. Part of the cloud of lighter material thrown from the original Earth would condense into the Moon and part would escape into space. The vast majority of the mass of denser material—including the iron

Looking around Our Solar System

from the cores of the asteroid and the original planet, very little of which would have been blown into space by the force of the collision—would come together to form the new Earth, pulling some of the lighter material back in along with it. Such a process would have been surprisingly rapid. Within fifteen minutes of the initial collision the soon-to-be-Moon would have been launched into space, and the two nuclei would have started to consolidate.

This theory explains many things about the Moon, particularly its size—we would expect the two new bodies to be very roughly the same size—and why its composition is so similar to that of the Earth's outer layers. As we saw, the Apollo missions had established that the Moon does have a core, apparently partially molten, but they were not able to determine whether that core contained any iron. The Lunar Prospector probe was likewise unable to confirm the existence of an iron core, but the measurements it made indicated that, if such a core existed, it could have a diameter of at most 600 kilometers. In other words, assuming the Moon has an iron core, its mass cannot be more than about 4 percent of the Moon's total mass. Scientists also accept the figure of 600 kilometers as the maximum diameter of any iron nucleus that could possibly have formed if the Moon had been created in a giant collision, given that it would primarily be lighter material, but very little iron, that would be hurled into space by the impact. Thus, the findings of the Lunar Prospector are in no way inconsistent with the asteroid theory. We cannot at present be certain that the Moon formed this way. But this theory best explains the available facts, and, at least to date, nobody has raised any fatal objections to it.

Desirable Real Estate?

Thirty years ago enterprising companies started to take reservations for the first commercial moon flight, but only now is commercial spaceflight finally getting underway. The first two paying astronauts have visited the International Space Station—at a cost of about $20 million for preparatory training and then the flight to the space station on a Russian Soyuz spacecraft—and a third is currently in training.[8] Although this does not mean that you or I will go into orbit, or even to the Moon, our children probably will. Events took an unexpected turn in September 2004 when Sir Richard Branston announced that he expected to fly a space liner with paying passengers in 2007 (see p. 235). It is still a long way from genuine space tourism, but the advent of the space tourist marks a necessary change in our approach to space travel. When the exploitation of space begins in earnest, ordinary people will need to be able to work in space with a minimum of special training, and spaceflight will need to become as routine (and as affordable) as transatlantic flights are today. When that happens, the Space Age will finally have arrived. The Moon has enormous potential, both as a scientific station and as an industrial base. The lunar vacuum creates a naturally sterile environment, and, unlike a space station, the Moon has gravity, which circumvents many of the medical problems associated with long periods of weightlessness and also offers a friendlier environment for construction work, in that the presence of at least some gravity allows structures to be securely anchored in terra firma.

With the overwhelming increase in the Earth's population, astronomical observatories are increasingly under siege. Optical observatories must contend with the spread of artificial light, which has already rendered some of the most famous of limited use. Similarly, radio telescopes must contend with the rapidly shrinking range of radio bands reserved for astronomy, and even these suffer more and more from man-made interference. The fact that orbiting telescopes are so difficult and expensive to gain access to severely curtails their value as a solution. (Just look at the problems the Hubble Space Telescope is suffering because of the need to repair and upgrade NASA's fleet of space shuttles.) The Moon, however, and particularly its far side, would be a perfect site for all types of astronomical research, what with fourteen-day nights and more than 3,000 kilometers of rock shielding radio telescopes from our noisy planet.

Plans already exist for an aggressive exploitation of the Moon's resources. The silicon-rich soil and uninterrupted daylight on the Moon's near side make it an ideal site for the generation of solar power. Solar cells could be set up that would produce huge amounts of cheap power that could be beamed to Earth. Moreover, the cells themselves could be constructed on the Moon in automated or semi-automated factories, using local materials. In addition, much of our mining for ore could at some point be transferred to the Moon. Lunar rocks can contain as much as 11 percent titanium, which raises the possibility that highly valuable processed ores could be produced on the Moon and then shipped back to the Earth. To accomplish this, a magnetic railway system, run by solar power, could be built to catapult ore off the Moon's surface. Although it is extremely expensive to send material *to* the Moon against the Earth's pull of gravity, the return journey is cheap and easy. (Compare the size of the lunar module, which had not only to land on the Moon but also take off again, with the giant Saturn V rocket required to handle the liftoff from the Earth.) It has even been suggested that the ecological battles of the late twenty-first century will be fought not to preserve the Earth's resources but to safeguard the last remaining virgin areas of the Moon.

It may be, then, that in the second half of this century the Moon will indeed become the most desirable real estate in the solar system.

SUGGESTIONS FOR FURTHER READING

Popular Books

Patrick Moore, *Patrick Moore on the Moon* (London: Cassell & Co., 2001).

> *In this popular book, first published in the 1940s as* Guide to the Moon, *the well-known English astronomer Sir Patrick Moore offers a complete guide to the Moon. Not all of Moore's opinions are standard; in fact, only in the past few years has the author accepted that the Moon's craters were the result of the impact of asteroids. Nonetheless, the book is still a wonderful guide to the Moon. For people who enjoy making observations with a telescope there are detailed maps and descriptions of the Moon's visible features. For armchair astronomers the book furnishes a wealth of detail that has been updated through the Lunar Prospector mission.*

Andrew Chaikin, *A Man on the Moon: The Voyages of the Apollo Astronauts* (Harmondsworth, Middlesex, Eng.: Penguin Books, 1994).

> *An admirably detailed description of the Apollo program from its initial planning to its conclusion. Based on interviews with all of the Apollo astronauts, on mission transcripts, and on NASA documentation, the book offers a fascinating inside look at the entire moon program, including the internal battles that went on, as well as facts about the missions that were previously either unknown or only known to a select few.*

More Advanced Reading

J. Kelly Beatty, Carolyn Collins Petersen, and Andrew Chaikin, eds., *The New Solar System*, 4th ed. (Cambridge, Mass.: Sky Publishing, 1998).

> *A collection of essays that includes an excellent article on the post-Apollo Moon, with splendid photographs and some very helpful diagrams that explain lunar geology in a very practical way. Beautifully written and prepared, and easily accessible to the advanced amateur astronomer.*

Grant H. Heiken, David T. Vaniaman, and Bevan M. French, eds., *Lunar Sourcebook—A User's Guide to the Moon* (Cambridge: Cambridge University Press, 1991).

A heavyweight volume that describes the post-Apollo Moon in great detail. Although not an especially easy read, the study is valuable for its very complete description of the results of the Apollo missions, including what was learned about the Moon's geology, chemistry, and physical characteristics, and about the various lunar resources.

On the Internet

The Moon
www.nineplanets.org/luna.html

A page from the Nine Planets Web site that provides a great deal of information, images, and links. One of the best places to start for those curious about the Moon.

The Moon
http://nssdc.gsfc.nasa.gov/planetary/planets/moonpage.html

A NASA site offering images, data, and bibliographic material, as well as links to the specific missions.

The Thirtieth Anniversary of the Apollo XI Mission
www.hq.nasa.gov/office/pao/History/ap11ann/introduction.htm

An official NASA site featuring information, interviews, and documents pertaining to the Apollo XI mission. Also available is an impressive collection of images from the mission, including some less well-known ones.

The Lunar Prospector Mission Page
http://lunar.arc.nasa.gov/

A complete guide to the Lunar Prospector mission, including a description of the experiments that were carried out and the results obtained.

CHAPTER 6

Is There Life on Mars?

O ver the past two hundred years, the way that the average astronomer would answer the question, "Is there life on Mars?"
has changed many times. Mars has captured our imagination
in a way that no other planet has. Books and films have repeatedly portrayed Mars as the aggressor. At the end of the nineteenth century, the
Martians of H. G. Wells's *War of the Worlds* destroyed southern England,
laying everything to waste as they advanced. In the mid-twentieth century, a radio dramatization of Wells's story precipitated a widespread
panic in the United States when thousands, if not millions, of Americans listening to the program concluded that Martians really had landed
and were taking over the country.

A century ago a significant minority of astronomers believed that
Mars might support intelligent life. Fifty years ago most astronomers
were sure that the planet was covered with dense vegetation. Thirty-five
years ago Mars had become a barren, sterile waste, not unlike the Moon.
Thirty years ago Mars was considered likely to be, if not teeming with
life, at least not lifeless. Twenty-five years ago the planet had returned

to being a sterile, barren landscape. Five years ago astronomers started to change their minds, yet again . . .

Why is it so hard to arrive at definitive answers about Mars? Why have we made so many wrong turns? Despite the resources we have devoted to studying the planet, the question that singer David Bowie asked in a song written at the time the Viking lander probes were being readied for their mission—"Is there life on Mars?"—remains unanswered. Why have we been unable to determine whether life exists on Mars and, if it does, what such life might be like?

Why Mars Is So Hard to Study

Until 1965, when the first successful Mars probe—NASA's Mariner IV—flew past the planet and sent back pictures and instrument readings, all our knowledge of Mars had been gathered by telescope. Astronomers had spent somewhere upward of two hundred years trying to examine the planet in detail with the best instruments at their disposal. Even so, because of the special difficulties it poses, Mars was nothing like as well studied as one might have thought.

Mars has been known to us since antiquity. Its red color has often invoked thoughts of blood, so in various cultures Mars has generally been named for the god of war. When the telescope was invented, Mars was one of the first bodies in the sky to be observed. The problem for astronomers is that Mars is a small planet. At 6,779.8 kilometers in diameter, it is just barely over half the size of the Earth, or a little less than double the Moon's diameter. Moreover, when Mars is at its very closest to Earth, which happens just once every fifteen or seventeen years, it is still at least 55.5 million kilometers away—almost 150 times as far away as the Moon. Thus, even at its nearest approach to Earth, it is a difficult object to observe through a telescope.[1]

But things are rarely so simple. Whereas the Earth's orbit is almost circular, the orbit of Mars is highly elliptical. In other words, Mars varies enormously in its distance from the Sun—from 207 million kilometers at its nearest point, or perihelion, to 250 million kilometers at its farthest, or aphelion. The point in time when Mars comes the closest to the Earth is called opposition. This is the moment when Mars is exactly on the opposite side of us from the Sun, so that the Sun, the Earth, and Mars line up—or, to put it another way, it is the moment when the Earth passes Mars in its orbit. Mars takes 697 days to orbit the Sun, which means it is in opposition to Earth approximately every twenty-two and a half months. Perihelion always occurs at the same position in Mars's orbit, and that position happens to correspond to the position of the Earth in late August, while Mars's position at aphelion corresponds to that of the Earth in February. Thus, if the Earth overtakes Mars in late August, when Mars is at perihelion, we have an excellent opposition because the two planets are as close together as they can be (as happened in 2003, when Mars and the Earth came nearer each other than they had for the past 58,000 years). In contrast, if opposition occurs in late February, with Mars at aphelion, the separation of the two planets is almost double what it is at perihelion. At the point of aphelion, Mars is nearly three hundred times as far away as the Moon and is therefore a tiny object in a telescope.

Unfortunately, things get worse. The Earth moves much faster in its orbit than Mars and thus both catches up to it quickly and rapidly leaves it behind. As a result, only for about a month on either side of opposition is Mars close enough to be relatively easy to study. (Actually, we are fortunate that over the next few years Mars will be at the most favorable part of its cycle of visibility, with the result that in October 2005 there will be a very good opposition, as there was in August 2003.) Thus, prior to the space probes, all our knowledge of Mars was based on observations made over about a two-month period every two years. In fact, only for some two months twice every fifteen years is Mars close enough to study in any real detail with a telescope. Is it surprising, then, that we have not been able to get to know the planet terribly well?

Canals and Controversies

The question of intelligent life on Mars might never have arisen were it not for a linguistic accident. Enter Giovanni Virginio Schiaparelli. In 1862, Schiaparelli became the director of Brera Observatory, in Milan. During the excellent opposition in the year 1877, he made a long series of observations of Mars with the 222-millimeter refractor at Brera. Up to then astronomers had been able to see the bright pole caps of the planet, which were assumed to be snowy. They had also observed various fixed dark markings on the surface of the planet that some suggested might be oceans, whereas the brighter regions between them suggested, correctly, deserts. They knew that Mars had an atmosphere but, by observing a star pass behind the planet and noting that it did not appear to fade before it disappeared behind the disk of the planet, William Herschel was able to demonstrate that the atmosphere of Mars must be very thin. In 1877, though, Schiaparelli observed something completely new and unexpected.

Crossing the deserts, from dark region to dark region, Schiaparelli spied thin, dark straight lines. On the assumption that the dark areas were seas and oceans, he called these new markings "channels," inasmuch as they connected "sea" to "sea" and "ocean" to "ocean." However, Schiaparelli, an Italian, used the word *canali*—"channels"—to describe them. The linguistically challenged (and presumably dictionaryless) anglophone astronomical community took *canali* to mean "canals," with the automatic supposition that they must be artificial. This was something that Schiaparelli had never intended and that would later have profound implications.

Schiaparelli made many observations of his "canali," both in 1877 and at the next opposition in 1879. He even observed a process he called gemination, in which a single canal would, between one night and the next, suddenly become a double structure with two canals in parallel, separated by a short distance. However, for many years, Schiaparelli was the only astronomer who reported seeing the canals. It was not until 1886 that two French astronomers at Nice Observatory, Henri Perrotin and Louis Thollon, confirmed Schiaparelli's canal network in all its glory. Although their observations were made during a rather unfavorable opposition, it was at this point that canals suddenly became fashionable. Had it not been for Perrotin and Thollon, Schiaparelli's canals might well have been swiftly forgotten.

Ironically, Schiaparelli never for a moment believed that the canals were artificial—at least not initially. It was only when others identified them as such that he became persuaded. The man who had most to do with popularizing the idea that the canals had been deliberately constructed was an American, Percival Lowell, who from the 1890s onward was a dedicated promoter of the theory. As Lowell depicted them, the canals crisscrossed the planet every which way, in a manner that has been likened to a diseased spider's web, since no healthy spider could possibly produce such a jumbled network of lines. One could believe that Schiaparelli's network of markings, although strange, might be natural, but Lowell's, if real, could only be artificial.[2] Lowell took things a step further and pictured a dying planet, populated by a highly advanced race, which had built a planetwide irrigation network in a desperate attempt to save their home.

Looking around Our Solar System

A huge amount of twentieth-century science fiction was based around the Martian canal system, which really captured the public imagination. But Lowell's ideas were more popular with the public than with astronomers. In fact, some astronomers openly ridiculed not only Lowell's observations but, even more so, his argument that the canals were evidence of intelligent life on the planet. Very few of them saw the fine, straight-line canals that he reported—although, at the same time, most if not all of them assumed that the canals had some basis in reality. In other words, astronomers were generally of the opinion that Lowell was seeing things that were really there but was interpreting them too imaginatively. Worse still for Lowell's reputation was the fact that he claimed to see similar networks of lines on other planets, such as Venus, that were virtually featureless to all other observers.

Until 1965, everyone believed that there was something on the surface of Mars that could explain the appearance of canals. Then Mariner IV sent back twenty-two rather heartbreaking images of a battered, crater-strewn landscape, with nothing visible anywhere that looked even remotely like a canal. Even though later probes such as Mariner IX photographed huge canyon systems, such as the Vallis Marineris (the Valley of the Mariners), these canyons did not coincide with any of the canals and so were not the explanation for them. Even today, the appearance of the canals is still a mystery. Nobody knows exactly what is being observed that causes people to see these straight lines, but they are still recorded by people viewing Mars through a telescope.

Vegetation?

Although the idea of intelligent Martians was discredited early in the twentieth century, it was still generally believed that Mars was teeming with life, if only in the form of lowly types of vegetation. As seen through the Hubble Space Telescope, or any other telescope large enough to reveal details of the planet's surface, Mars has three main colors. Much of the planet is orange-red; this is the color of the deserts and is due to the quantity of iron oxide in the soil. The pole caps are white, or bluish-white, as are the clouds, particularly when they appear toward the edge of the planet's disk. Finally, there are the dark regions, with their characteristic blue-green color. Although the blue-green color is not as intense as some books suggest, it is clearly visible. Some years back, I was fortunate enough to be able to look at Mars through the 4.2-meter William Herschel Telescope on the island of La Palma.[3] The deep blue-green color of the dark areas was very obvious, although through a small telescope they will always appear grey.

The color of these markings suggested vegetation, as did certain features of their behavior. In spring, as the pole cap receded, sending moisture into the atmosphere, many observers reported a wave of darkening spreading away from the pole, during which the markings darkened and their color appeared to intensify. From early in the twentieth century it was generally accepted that this phenomenon was produced by the regeneration of vegetation after the winterkill. Moreover, these dark regions always reappeared, essentially unchanged, even after the immense global dust storms that sometimes covered the planet's entire disk. The fact that they were not hidden by windblown dust was taken as proof that these areas must be made up of vegetation, which would grow up through the covering of dust, just as plants do on Earth when they have been covered by ash or dirt.

Although a few astronomers were unconvinced, most believed that the evidence that the dark areas were due to some form of hardy vegetation, probably moss or lichens, was fairly conclusive. Here too, though, the first Mars probes supplied a number of unpleasant surprises. Up to 1965, the majority of estimates of the atmospheric pressure on the Martian surface gave a value of about 100 millibars. This is one-tenth of the pressure on the Earth's surface, equivalent to the pressure at an altitude of 17 kilometers, or double the height of Mount Everest. When Mariner IV flew past Mars, the attenuation of its radio signals as it passed behind the planet was used to provide an estimate of the atmospheric pressure. The results were as unexpected as they were disappointing. The surface pressure of Mars turned out to be more than ten times lower than had been anticipated—between 4 and 7 millibars. What is more, most of the atmosphere seemed to consist of carbon dioxide. This last discovery seemed to rule out the possibility of pole caps made of ordinary ice and suggested instead that they were composed of dry ice—frozen carbon dioxide.

It now seemed almost impossible that the new Mars—cratered, desolate, with a desperately thin, totally dry atmosphere—could be an abode for life. Then Mariners VI and VII, which flew past the planet on July 31 and August 4, 1969, just days after the triumphant return of the Apollo XI astronauts, made another disconcerting discovery. The two probes photographed Mars as they were approaching the planet, unlike Mariner IV, which had only taken photographs at

Looking around Our Solar System

its closest approach. The first images, taken from over 800,000 kilometers away, showed the dark markings very clearly. As the two probes approached the planet, however, these markings just faded away, and in the images taken from nearest the planet they could not be discerned at all. Whatever they were, if they could be seen only from a long way away, it was obvious that they were not tracts of vegetation. It is now thought that the dark markings must be regions that are blown at least partially clear of dust.[4] Prior to the Mariner probes, well-known dark regions, such as Syrtis Major and Mare Acidalium, were always assumed to be low-lying areas, possibly even ancient seabeds. We now know, however, that they are windblown plateaus, while Hellas, previously assumed to be at high altitude, is in fact a huge crater and the deepest basin on Mars. What is certain, though, is that neither the Syrtis Major nor any of the other prominent markings seen from Earth are evidence of vegetation.

Rivers and Volcanoes

In late 1969, then, the situation for Mars researchers was depressing. Three probes had flown by the planet, Mariner IV, Mariner VI, and Mariner VII, returning 22, 75, and 126 images of the surface, respectively. (Of the Mariner VI images, 25 were taken at closest approach, as were 33 of the Mariner VII images. The rest were far-encounter images taken as the probes approached.) About 20 percent of Mars had been photographed in detail, and there appeared to be nothing to indicate that the rest of the planet would not follow the same pattern of heavily cratered, chaotic terrain. There seemed little point in sending more probes to such an obviously dead planet. Had the twin Mariner VIII and IX probes not already been so far along, it is likely that they would have been canceled. Fortunately, they were not.

Mariner VIII fell into the sea on launch, after its second stage failed to ignite. Much then depended on Mariner IX, launched three weeks later, on May 30, 1971. Had Mariner IX failed as well, it is unlikely that NASA would have attempted further Mars probes. As it was, despite a perfect launch and a flawless flight, Mother Nature, in its Martian guise, had some unpleasant surprises in store for the probe. The 1971 opposition of Mars was an extremely good one. For a month after opposition, with Mariner IX and two Russian probes—Mars 2 and Mars 3—closing in on the planet, all was well. Then, in mid-September, a large dust storm developed. For a few days it appeared that it would not be too serious; however, suddenly, at the end of September, the storm spread until it covered the entire globe. When Mariner IX began orbiting Mars, on November 13, almost nothing could be seen on the planet. Over the next three weeks Mars 2 and Mars 3 also arrived and were similarly blinded.[5] The only details that could be seen in the Mariner images were four small spots, evidently the tops of high mountains, three of which were nicknamed, with devastating originality, North Spot, South Spot, and Middle Spot.

It is fortunate that Mariner IX, which was only intended to function for ninety days, in fact continued to operate for almost an entire year until it finally fell silent. After six frustrating weeks the storm subsided, and the probe could finally start its task of mapping the planet. The three spots were found to be giant volcanoes that are now known as Ascraeus Mons, Pavonis Mons, and Arsia Mons. The fourth spot, though, was the "king of the mountains." Olympus Mons (Mount Olympus), as it known, is the highest mountain in the solar system, with a height of some 21 kilometers, three times the height of Mount Everest.[6] Mariner IX also photographed the great canyon systems. One of these, Vallis Marineris, would extend across the United States almost from coast to coast and, if filled with water, as most experts believe it was at one time, would have a flow equivalent to the entire Amazon basin. At the other extreme we have the narrow sinuous valleys that look like dry riverbeds, such as Kasei Vallis and the amazingly complex canyon system of Noctis Labyrinthus (the Labyrinth of Night) that, because of its peculiar shape, is nicknamed "the Chandelier." Mariner IX also showed that, contrary to previous belief, the pole caps appeared to be largely composed of ice.

Suddenly Mars had turned into an active world again. There was plentiful water in the pole caps, ample evidence of flowing water around the planet, and huge volcanoes to put gas into the atmosphere. Even though some researchers were unhappy because they had hoped to discover a large amount of frozen carbon dioxide in the poles that could potentially provide a much denser atmosphere for the planet, all in all Mariner IX gave a huge boost to the prospects for life. Now, it was not so much a question of "if," but rather "how much?" Unfortunately, things were about to go horribly wrong again.

Where Are the Bodies?

The Viking landers were intended to be NASA's bicentennial present to the American nation. Viking 1 was scheduled to land on July 4, 1976, but the landing site originally chosen turned out to be too rough and rocky. An extensive search was required to find a suitable alternative, which delayed the actual landing until July 20. It and its sister probe, Viking 2, carried television cameras, meteorological instruments, atmospheric analysis equipment, soil analysis equipment, a seismometer, and a small chemical laboratory designed to detect signs of life in the soil. Shortly before the landing the BBC asked Carl Sagan his opinion of the chances of finding life. His reply reflected the bullish optimism of the day: "If there is one problem that we are likely to have with Mars, it's that 'we didn't guess the organisms right.' That we poor scientists of the last half of the twentieth century were incapable of second-guessing an organism that found a way of doing something that we couldn't think of." In other words, as Sagan saw it, the problem was not so much whether or not life existed. Rather, the danger was that this life might be too alien for us to detect unambiguously. Perhaps Martian life forms would have a biochemistry so different from that of earthly organisms that the experiments to be carried out would not work successfully with them. Interestingly, Sagan did not even rule out the possibility that there might be larger Martian life forms and even things analogous to animals. "I have this sort of nightmare that every morning, at sunrise, we see the area around the Viking lander covered with footprints," he said. "However, the TV cameras will be able to detect them—if there are big Martian beasts."

As has happened so many times with Mars, the planet had a major surprise in store. The Viking landers carried three simple but extremely clever experiments aimed at detecting life in the Martian soil: pyrolytic release, labeled release, and gas exchange. All three involved doing something to the Martian soil to see how it would react. The philosophy of the experiments was simple: a positive result would indicate that microbes in the soil were reacting.

Pyrolytic release was the simplest, a sort of brute-force approach. Radioactive carbon dioxide gas was added to a chamber containing a sample of soil gathered by the lander's robot arm. After eleven days, during which the chamber was lit to simulate sunlight, it was heated violently to 627°C. The idea was that any life-forms in the soil would absorb the radioactive carbon dioxide, which would then be released when the soil was heated violently. The release of radioactive gas would indicate that organisms had used the radioactive carbon dioxide in the same way that Earth's plants use ordinary carbon dioxide in photosynthesis.

When the soil was heated, it *did* release some radioactive carbon dioxide. This encouraging result was, however, soon cast into doubt. As good scientists do, the Viking team checked their results. They ran the same test on a sample of soil that had first been sterilized by heating it. To their surprise, and their disappointment, the sterilized sample yielded the same positive reaction, which demonstrated that the positive result was not brought about by any organisms in the soil. Scientists now think that the positive result is *probably* due to the unusual iron compounds—peroxides and superoxides—found in the Martian soil, although some questions remain. All the same, the results of the pyrolytic release experiment are not now interpreted as in any way indicative of Martian life.

Labeled release was an altogether more gentle method, intended to determine whether organisms in the soil sample would react if wetted with water. The soil sample was moistened with a nutrient solution that included carbon-14, the same radioactive isotope of carbon used in the previous experiment. The assumption was that organisms would absorb the carbon-14 and then "breathe it out" as gas. The release of radioactive gas from the soil would thus be regarded as almost conclusive evidence of biological activity in the soil. The results from this experiment are particularly complicated to understand and are still being debated nearly thirty years later. There was an initial rapid and massive release of radioactive gas. Something in the soil was digesting the organic material in the nutrient solution. The release of radiation soon started to level out, however. After seven days, a second dose of radioactive nutrient was added. What happened then was totally unexpected: the amount of radioactive carbon released dropped suddenly by about 30 percent. A third dose of radioactive nutrient produced no reaction whatsoever. Was Carl Sagan correct and this strange result due to the peculiarities of Martian organisms? Or was it brought about by odd chemical reactions in the soil?

The experiment was then run with sterilized samples, again with no reaction. However, this could have been because the chemicals in the soil had themselves been destroyed by the high temperatures—between 200°C and 500°C—used to sterilize the soil. A new sterilization technique was therefore tried that should, in principle, have decided the question. A soil sample was "cold sterilized." The soil

Looking around Our Solar System

was heated to only 50°C, a temperature high enough to kill most organisms but not so high that it would break up any chemicals in the Martian soil. But when the experiment was repeated with this soil sample, it barely reacted. The sterilization had destroyed, or killed, whatever it was in the soil that had initially reacted. All in all, the results of the labeled release experiment were sufficiently peculiar that they left the scientists completely bewildered. It is fair to say, though, that this experiment can be interpreted as evidence either for unusual Martian organisms or for unusual Martian soil chemistry.

The third experiment, gas exchange, used a nonradioactive nutrient solution that the scientists called "chicken soup"—although it had no chicken in it and was intended only to be appetizing to Martian microbes, not to the scientists who concocted it. The idea was to add this nutrient solution to the soil and then see what gases the soil "breathed out" in response. Initially, all that was done was to fill a cup located in the chamber with chicken soup, but not to wet the soil itself. Mars is a very dry world, so once the cup was filled with liquid, the atmosphere inside the chamber would become humid. After a week, the level of soup was increased, and a few drops were allowed to spill onto the soil.

Once again the results were totally unanticipated. As the soil was humidified it produced a huge burst of oxygen. However, when the soil was moistened with the soup, the only reaction was the release of a little carbon dioxide. These results suggested that the soil contained an extremely powerful oxidizing agent—something like strong household bleach. It was this Martian bleach that was initially causing the positive results—the burst of oxygen—by breaking up the nutrient solution. However, the bleach itself later broke down when wetted.

On the whole, then, the results of the biology experiments were very confusing. At an early stage in the analysis, Gerald Soffen, project scientist for the Viking missions, famously summed up the situation by saying: "All the signs suggest that life exists on Mars, but we can't find any bodies." After considering all the results, the biology team's carefully worded comment was that "the Viking results do not permit any final conclusion about the presence of life on Mars."[7]

The results of the Viking experiments would have remained ambiguous were it not for a check that the scientists had obtained. If there are organisms in the soil, it must be possible to find their bodies—the organic compounds that all life contains. The Viking landers carried a small apparatus called a gas chromatograph mass spectrometer. A soil sample would be heated in stages to a temperature of 500°C and the gases released then analyzed to measure the mass of their molecules. The molecules could then be identified on the basis of the masses detected.

Everyone expected the soil to be teeming with organic compounds, but it was not. In fact, no organic compounds were detected at all. A simple explanation might have been that the apparatus was not working. However, an in-flight test of the spectrometer's oven had revealed that it was still contaminated by the acetone used to clean it before the launch. The spectrograph detected the acetone contamination perfectly as the oven boiled it away, which indicated that everything was working correctly. The only possible conclusion was that something in the Martian environment destroyed organic compounds rapidly. Quite apart from the oxidizing agent in the soil, the surface of Mars is bathed in lethal ultraviolet and cosmic radiation. An extended stay on the surface would be extremely unhealthy for any astronaut. The presence of all this radiation has given the impression that perhaps the surface of Mars is constantly sterilized and that no life forms *can* exist there.

In a final attempt to obtain a positive result, the Viking lander's robotic arm moved a rock so that soil could be gathered from underneath, where it would have been protected from radiation. This soil sample, though, was just as sterile as those from the surface. Writing in 1990, Soffen concluded: "I have devoted 20 years to thinking about this question. In my role as the Viking project scientist during the primary mission, I began with an optimistic view of the chances for life on Mars. I now believe that it is very unlikely. But one doubt lingers: we have not explored the planet's polar regions . . . the oxidizing agents might be absent—due to the presence of water."[8]

The Return of Life on Mars

In the mid-1990s, after twenty years of imagining Mars as a sterile planet, scientists again became excited by the prospects that the planet could have life. The reason was not something found in the Martian deserts, however. It was instead a find in the frozen wastes of Antarctica—a small stone called ALH84001 that has caused a scientific furor. "ALH" stands for Allan Hills, the place in Antarctica where the rock was found, while "84" is the year of its discovery (1984), and "001" means that it was the first meteorite of the year found at that site.

Looking around Our Solar System

Several years after its discovery, this meteorite was recognized as being of a very unusual type called a shergottite. The meteorites of this type all seem to have come from Mars, and they have a distinctive composition in comparison to meteorites that are known to be pieces of asteroids. The most convincing evidence that ALH84001 and a few other meteorites originated on Mars is the fact that the composition of the gases trapped inside tiny glass bubbles within the meteorite is identical to that of the Martian atmosphere. The Earth's atmosphere contains about 1 percent of a gaseous element called argon, almost all of which is argon 40, or ^{40}Ar, to give it its chemical symbol. However, there are smaller amounts of two lighter isotopes of argon, ^{36}Ar and ^{38}Ar. The Earth's atmosphere contains about 300 times more ^{40}Ar than ^{36}Ar, whereas Mars's atmosphere has 3,100 times more. These ratios and others like it—such as that among the different isotopes of the rare gas xenon—are a fingerprint of the atmosphere of each planet and a source of information about the origin and history of that atmosphere. The gas trapped inside the glass bubbles in ALH84001 has the characteristic fingerprint of Mars, totally unlike that of any other known body in the solar system.

Scientists are confident that ALH84001 was blown out of Mars by the impact of a large asteroid some sixteen million years ago. What is important about ALH84001 is that it contains organic compounds—and not just small quantities of them. In fact, some 0.1 percent of the microscopic carbon-rich grains or "globules" within the meteorite is composed of relatively massive, organic molecules. What is more, the grains are embedded within the meteorite, and the amount of organic compound increases as one moves deeper within the grains, indicating that these organic molecules are not the result of terrestrial contamination.

So we have a rock that came from Mars—that much is not in dispute—and that, unlike the soil analyzed by the Viking landers, is rich in organic compounds. Had the NASA announcement said only this much, it would still have been astonishing news, all the more so because teams of scientists analyzing other Martian meteorites subsequently discovered something similar. However, the most surprising, and controversial part, of NASA's announcement was the suggestion that there was evidence of the action of Martian bacteria in the meteorite. This conclusion was based on three individual pieces of evidence, none of which is conclusive on its own but which, taken together, are at least suggestive:

1. The organic compounds are found on the surfaces of fractures within the meteorite. When microbes on Earth die, they break down into molecules known as PAHs (polycyclic aromatic hydrocarbons), like those found in the meteorite. Those found in the meteorite, however, appear to be different from the PAHs found in typical contaminants on Earth, such as dust, that might have gotten into the meteorite.

2. Many of the carbonate globules have cores containing tiny crystals of iron minerals, particularly magnetite and iron sulfide. These are similar to the crystals produced by certain rock-eating bacteria on Earth.

3. Electron microscopy has revealed the presence of tiny wormlike structures that may be the fossil remnants of tiny (20 to 100 nanometer) bacteria. If so, they are a hundred times smaller than any bacteria microfossils found on Earth.

Some scientists feel that these findings point to the possibility of primitive bacterial life on Mars, at least in the distant past. But such conclusions are highly controversial. If we accept that what was discovered in the rock is indeed the product of Martian bacteria, the history of ALH84001 would be something like this:

- The original rock solidified within Mars about 4.5 billion years ago, about one hundred million years after the formation of the planet.
- Between 3.6 and 4 billion years ago the rock was cracked, probably by the impact of meteorites. Water then entered the cracks, depositing carbonate minerals and allowing primitive bacteria to live in the fractures.
- About 3.6 billion years ago, the bacteria and their remains became fossilized in the fractures.
- Then, roughly 16 million years ago, a large meteorite struck Mars, dislodging a large chunk of the rock and ejecting it into space.
- Somewhere around 13,000 years ago, the meteorite landed in Antarctica.

The scientists who doubt that the meteorite really shows evidence of fossil life base their arguments on four possible problems. First, they argue that the liquid that flowed through the cracks in the meteorite may not have been water. If the rock was buried deep underground, it is possible that the substance involved was carbon dioxide liquefied under enormous pressure. It is hard to imagine any kind of life that could live in liquid carbon dioxide; therefore, the organic compounds that formed would almost certainly not have been produced by living organisms. Second, some scientists argue that the organic compounds appear to have formed at very high temperatures, perhaps as much as 600°C—temperatures at which no living organism could possibly survive. If the organic compounds did indeed form in such an extremely hot environment, it is simply not conceivable that they came from living organisms.

The third argument pertains to the tiny crystals of iron minerals found in the core of many of the carbonate compounds. Scientists point out that only a few types of terrestrial bacteria produce magnetite crystals, and it would be a highly unusual coincidence if such bacteria were to be duplicated on Mars. The mineral structures may therefore be completely natural, although they have been the subject of numerous studies and counterstudies, with some scientists claiming that the crystals have a unique structure reminiscent of structures produced by bacteria on Earth and others arguing that the crystals are a product of nature. Similarly, the long, wormlike structures have been interpreted not as the fossilized remains of tiny bacteria but as natural magnetite whiskers formed at high temperatures. Finally, some scientists argue that the organic compounds found inside the meteorite are the result of terrestrial contamination, caused by meltwater entering the rock after it fell in Antarctica.

As matters presently stand, the scientific community is completely split on the issue of whether the meteorite ALH84001 shows evidence of life. Every few months one reads in the press of new discoveries that "prove" that the organic compounds have a natural origin or that "demonstrate" that that they were probably produced by Martian organisms. In fact, neither side has proved its case yet. Some two years after NASA made its announcement I surveyed the articles about the meteorite that had since been presented in learned journals and at conferences. Exactly a third supported the theory that ALH84001 showed evidence of Martian organisms, exactly a third opposed the theory, and exactly a third made no judgment either way!

In February 2001 a NASA-supported group claimed to have obtained definitive proof that ALH84001 shows the effects of biological activity. They produced a series of studies that revealed similarities between terrestrial rock-eating bacteria and the structures in ALH84001. The structure of the magnetite crystals in ALH84001 was found to be identical to those produced by the MV-1 bacterium, while the whiskerlike structures have also been seen in studies of the fossilization of terrestrial bacteria. The most recent piece of evidence is the alignment of the magnetite crystals in chains, something also observed in grains produced by bacteria on Earth. These features were presented as proof that the magnetite crystals in ALH84001 were produced by Martian biological activity. Interestingly, on the release of the latest news of ALH84001 another NASA group announced that similar structures had been found in two other Martian meteorites: Nakhla and Shergoty. Whereas ALH84001 is a very ancient rock, these two meteorites are much more recent—1,300 million and 175 million years old, respectively. If the carbonate grains truly are in part biological in origin, this would imply that life survived on Mars until quite recently, perhaps during as much as 96 percent of the present age of the planet. As before, however, all these findings have been received with great skepticism by some experts and with guarded interest by others. The case is still far from closed.

Recent Discoveries

The past few years have witnessed a host of important discoveries about Mars. Scientists have known since 1971, when Mariner IX orbited Mars, that the planet has an active weather system, with clouds, frontal systems, and even the odd swirl of cloud that marks a cyclone or an anticyclone. But they confidently assumed that Martian clouds were very thin. Mars Global Surveyor—launched in 1996 and, in 2004, still in orbit around Mars, taking high-resolution images of the surface—has since made direct measurements of the thickness of the clouds by shining a laser beam down to the planet's surface. To the considerable surprise of scientists, the clouds inside some of the deep canyons are far more dense and opaque than was ever believed possible. Similarly, laser-ranging measurements of some of the canyons have shown that they are deeper and steeper than was thought. Not only was the volume of water in them up to three times greater than previously estimated, but that water flowed at very high speed, up to 130 meters per second.

Scientists now recognize that Mars must have had huge quantities of water in the past as well as a much milder climate. There is increasing evidence that an immense ocean once surrounded much of the northern hemisphere of Mars, wrapping around the north pole of the planet. As Mariner IX demonstrated, much of this hemisphere is lower in altitude than the southern hemisphere, which prompted some researchers to speculate that perhaps part of this hemisphere is a huge crater created by some nearly planet-splitting impact in the remote past. Mars Global Surveyor has revealed that this region is flat and lies as much as 6 kilometers below the mean level of the planet. If one fills this region with water to the notional Martian "sea level," forming what has been dubbed the Polar Ocean, or Oceanus Borealis, one finds that the Vallis Marineris drains into it in Chryse, the desert region where Viking I landed. Chryse itself would be a large bay in this circumpolar ocean. If indeed Mars had an ocean and great rivers in the past, then the atmospheric pressure must have been much higher, and the greenhouse effect, which is now very weak on Mars, must have been much stronger, allowing liquid water to exist without freezing or boiling away.

One of the tasks of current and future explorations of Mars is to try to work out the history of the atmosphere and of water on the surface of the planet. How dense was the atmosphere? For how long was the climate much warmer than it is now? Why did the atmosphere disappear? Has the climate been periodic, as has been proposed, with regular cycles of warm conditions and ice ages? Were episodes of violent vulcanism, which released huge quantities of gases into the atmosphere, responsible for the relatively warm and wet periods? Where is the water now? Is it trapped below the surface as ice?

Perhaps the most surprising recent announcement has been that water may still flow on the planet. Mars Global Explorer sent back images of a Martian cliff marked with streaks that look identical to the courses of mountain streams on the Earth. In this "weeping terrain," as it has been called, the pressure of rock has probably melted ice under the surface, leading to occasional bursts of water out of the cliffside. This theory has its dissenters, though, with some scientists arguing that the streaks were produced by liquefied carbon dioxide, under enormous pressure from hundreds of meters of rock on top. Nobody knows whether water (if that is what it was) last flowed yesterday, a century ago, or ten thousand years ago, but we know that, in geological terms, it was recently. One of the aims of the current European Mars Express mission is to use a special decameter radar, called MARSIS, to search for underground lakes. This radar technique, used in the cold war to search for underground missile silos, uses radar with a wavelength of tens of meters, instead of the usual few centimeters. Part of the radar signal is able to penetrate the planet's surface and is reflected back from underground layers of material. Computer analysis can then produce a map of the subsurface from the signals received. Unfortunately, as this book goes to press, the deployment of this radar has been delayed for yet a second time, owing to a technical problem, and is now currently planned for May 2005. Results are therefore not yet available—and in any case the search for subterranean lakes is liable to be a long one. Clearly, though, the discovery of underground galleries full of water would have profound implications for the prospects of finding life on the planet and for future human settlement of the planet.

Adding It All Up

As I write, the Mars Exploration rovers, Spirit and Opportunity, which landed on Mars in January 2004, are still roaming around searching for evidence of water at two points on opposite sides of the planet. Within weeks of landing the two rovers had, for the first time, discovered both hydrated minerals—that is, minerals with water trapped in their crystals—and carbonates such as those discovered in the meteorite ALH84001. And yet the initial results have seemed ambiguous. The Spirit rover has detected the mineral olivine in the soil of Gusev crater, which is thought to have once held a large lake that at some point breached the wall of the crater and drained away, carving the spectacular gorge now visible to the southeast of the crater. But olivine—a volcanic mineral often found in lava—breaks down into other minerals in the presence of water, which would seem to argue against the lake theory. However, the Spirit also found strong indications that water had indeed once been present: hydrated minerals, hematite, and salts that appear to have been deposited as saline water evaporated. One possibility, then, is that after Gusev crater drained, what water was left gradually dried up, producing the salts, and then much of the bed of the lake was eventually covered by a lava flow, which would account for the olivine. In the meanwhile, the Opportunity, which has been exploring the plain known as Meridiani Terra, has located large concentrations of the iron mineral hematite, which is often formed in the presence of water. It is still not obvious, though, whether the hematite is of the type associated with water or the type sometimes produced by volcanic activity, although the evidence favors the former and suggests that the Meridiani landing site was also once under water. More recently, the Spirit, which is now climbing high up into the peaks of the Columbia Hills, above the plains of Gusev crater, has further detected substantial traces of minerals that have been modified by water.

Even though a great deal of work remains to be done analyzing the data that continue to come in from Mars, it seems safe to say that we now have solid evidence of the past presence of liquid water on the Martian surface. Not just that, but there appears to have been water in considerable quantities—certainly larger than had been expected. All the same, the evidence we have assembled concerning the possible existence of life on Mars clearly has no easy interpretation. The conditions on the surface were much more favorable for life in the past, and certain of the data collected by probes suggest that Mars may be emerging from a global ice age and the temperature may be systematically rising on the planet. We know that, at the time when water was flowing over the surface of the planet, the atmospheric pressure was many times what it is today and the temperature far higher, and so it is not hard to imagine that life may at some point have at least taken its first steps on Mars. Some experts even speculate that life on Earth arrived by way of meteorites from Mars. But the critical question has to do with how long these warm and wet conditions lasted. Did Mars have a long wet past or a short wet past, in the argot of the areologists (as people who study Mars are termed)? If it were the latter, perhaps life never managed to get started before the climate changed and the atmosphere was either dissipated or frozen out. If it were the former, though, it would be disappointing to discover that life never actually developed on Mars.

My own feeling is that evidence increasingly points to the conclusion that Mars was a hospitable world for much longer than was previously thought. In particular, if the Oceanus Borealis, the northern polar ocean, really did exist for hundreds of millions of years, very probably life did get underway. From there, however, it is a very large jump to the idea that life has survived until today—and yet one thing that we have learned about our world is that life is incredibly tenacious. Once it gets a toehold, it hangs on. Moreover, life manages to exist in the most extreme environments. Organisms have been found that live in nuclear reactors and in the near-boiling water of hot springs. They live inside rocks in the Antarctic and at the greatest depths of our oceans. They live high in the atmosphere, and they live in Spain's Rio Tinto—a river whose water is colored deep red by enormous concentrations of poisonous heavy metals. Bacteria were even found inside the camera of the Surveyor moon probe that the Apollo XII astronauts visited, which were still viable after two years on the Moon's surface. If there is any possible way that life can survive by adapting to an environment, it will find it.

That said, conditions on the surface of Mars may have been so thoroughly inhospitable in the past that even the toughest and most tenacious organisms were finally destroyed. The Viking landers found that at two locations on the planet's surface, separated by a considerable distance, the soil was totally sterile, and all the evidence points to the fact that something in the soil actively destroys organic compounds. It is unlikely that there will be any great difference in the conditions at other points around Mars—but then the soil is not the only potential habitat for life. There is evidence to suggest that water still flows beneath the surface of Mars, and water seems to destroy the oxidizing chemicals in the soil. Some scientists have therefore suggested that the icy Martian pole caps, where frozen water is abundant, are the most likely place to find life, even though the temperatures at the Martian poles are appallingly low. This possibility will be explored in the Mars Reconnaissance Orbiter mission, planned for launch on August 10, 2005.

It is unfortunate that the British Beagle 2 probe has been lost, for it was to have carried out the first-ever check of the Viking results. It was equipped with an instrument called "the Mole" that could burrow under the surface and take soil samples from beneath rocks and with a sensitive mass spectrograph that could detect extremely tiny quantities of organic compounds in the soil. The spectrograph was also capable of distinguishing between organic compounds produced by life and those produced by chemical processes. With the loss of Beagle 2 it will not be until 2007, at the earliest, that we will have our first opportunity to determine whether, as some experts have suggested, the Viking probes were not sensitive enough to detect the tiny concentrations of organic material that would be present in the soil if there were only a few hundreds, or even a few thousands, of microbes per cubic centimeter. Some argue that even very low concentrations of organisms could have produced the initial positive result of the labeled release experiment carried out by the Viking landers.

However, despite the wealth of information gathered by the robot cars and by orbiting space probes, what is arguably the most exciting discovery concerning life on Mars originated with telescopes here on Earth. Both the NASA Infrared Telescope Facility (IRTF) on Mauna Kea, in Hawai'i, and the Gemini South Telescope, in Chile, have detected evidence of methane in the Martian atmosphere, findings that have now been confirmed by the Mars Express space probe. Methane (CH_4) is a very unstable gas that is rapidly broken down in the Martian atmosphere, so in order for the gas to be present, it must be replenished constantly. On Earth, there are two sources of atmospheric methane. One is bacteria—as cows digest grass, for example, the action of bacteria in their stomachs generates large amounts of methane—and the other is volcanoes. Although Mars boasts some impressive volcanoes, as yet we have no proof that they are active, and, even if they do keep up some residual activity, whether they could emit enough methane to account for the quantity detected is an item of debate. It is at least possible, however, that even though the Viking landers were unable to locate any traces of organic material, somewhere on Mars are bacteria, in significant numbers, that are producing methane gas. The definitive test will be to measure the distribution of methane around the planet. If it proves to be concentrated around the great volcanoes, then it is probably volcanic in origin, whereas if it is fairly evenly distributed all over the planet, then it is almost certainly biological. By the end of 2005 we may possess the strongest proof yet that life exists on the surface of Mars.

Not only has methane been detected but possibly also ammonia. An instrument called a planetary Fourier spectrometer, carried by the Mars Express probe, has tentatively detected a 10-micron line of ammonia in the Martian atmosphere—although, if the data collected by the spectrometer are the source of much excitement, they are also proving far more difficult to analyze than anticipated. Ammonia (NH_3) is of particular interest because, like methane, it can be either biological or volcanic in origin and is also extremely short-lived (it is broken down within hours by solar radiation), with the result that it must be constantly replenished. Moreover, ammonia contains nitrogen, which is very rare on Mars. Because nitrogen is an essential ingredient of proteins, life is not possible without it. While it is not inconceivable that ammonia could have been released by the airbags of the Beagle 2 probe, it does not seem to be concentrated around the probe's landing site. If we can confirm that ammonia is present in the Martian atmosphere, this will lend strong support to the idea that huge numbers of active microbes exist on the planet.

But, as with so many things we have discovered in our exploration of Mars, the detection of methane and possibly of ammonia can be interpreted in two ways. It may simply indicate that Mars still has residual vulcanism and that the amount of gas released from the volcanoes is far greater than had been imagined. Recent results from the Mars Express probe have suggested that the giant volcanoes on the Tharsis ridge—Olympus Mons, Pavonis Mons, Ascraeus Mons, and Arsia Mons—could have been active as little as one million years ago. Scientists were able to arrive at this estimate by looking at extremely high-resolution images of the volcanoes and counting the number of small craters that are visible. The fewer the craters, the younger the ground must be. The fact that we know the approximate rate at which craters are formed by asteroid bombardment then allows us to calculate the date of the most recent volcanic activity. As Mars Express project scientist Agustin Chicarro notes, it would be an unusual circumstance if these giant volcanoes, which were probably active for tens of millions, if not hundreds of millions, of years, had suddenly ceased to be active only a few million years ago. Then again, other scientists have cast doubt on the conclusions drawn from the Mars Express, which are indeed still tentative, and have reminded us that a host of other Mars probes have yet to detect any signs of active volcanoes and that small craters can be obliterated by erosion resulting from dust storms, making a surface look younger than it really is. Similarly, some scientists argue that the amount of methane detected in the atmosphere far exceeds what could be explained by volcanic outgassing and that such quantities of the gas can only be explained by the presence of extensive microbial activity. My own view is that, on balance, the present evidence—while far from conclusive either way—points more to the existence of life than to active volcanoes. As matters currently stand, though, we may still be years from learning the truth.

If, however, there are microorganisms on Mars, then where might they be found, given that the two Viking landers were unable to locate them? As we have seen, some scientists suggest that we investigate the Martian poles. But there are two other places to look for life that may actually be more promising than the poles. One is the bottom of the Vallis Marineris. There, evidence of water exists in the form of morning mists. But although it is often said that the atmosphere is much warmer and thicker deep in the Vallis Marineris than at the Viking landing sites, this is not true. Viking I's landing site in Chryse was 3.5 kilometers below the Martian "sea level." While the deepest parts of Vallis Marineris, in the east, are 4.5 kilometers below sea level—deep enough to have a slightly thicker and warmer atmosphere than Chryse—most of the valley floor is actually at a higher altitude than the Viking I landing site.

If we really want a promising site for life, though, I would suggest looking at the great volcanoes. If some volcanic activity is still present, we will find hot, or at least warm, rocks underneath the surface. There will also be water, because we know that the Vallis Marineris flowed downhill from these volcanoes to the

Looking around Our Solar System

Oceanus Borealis. There, below the surface and therefore protected from harmful radiation, we will find a habitat that may well be ideal for any surviving organisms, probably in the form of very simple rock-eating bacteria. Finding them, though, is likely to be a difficult feat to accomplish unless these Martian bacteria turn out to be unexpectedly plentiful.

SUGGESTIONS FOR FURTHER READING

Classic Science Fiction

H. G. Wells, *The War of the Worlds* (London: Pan Books, 1979 [1898]).

> *The classic tale about a Martian invasion of the Earth. Although after more than a century it is inevitably dated, it is still very worth reading.*

Arthur C. Clarke, *The Sands of Mars* (London: Sidgewick and Jackson, 1982 [1951]).

> *A fascinating story about Mars as it might have been, based on the best knowledge available in the 1950s. Remarkably, even now, more than fifty years later much of the story is still valid.*

Kim Stanley Robinson, *Red Mars* (London: HarperCollins, 1992).

> *The first and by far the best of a trilogy, this book is an extraordinary document, full of scientific detail. An enjoyable way to learn about Mars and about how it might be explored and colonized in the future.*

Ben Bova, *Mars* (London: Hodder and Stoughton, 1992).

> *A somewhat different view of Mars from Kim Stanley Robinson's, with less hard science, and in some cases events have already moved ahead of Bova's account. All the same, still a fascinating and informative read.*

Popular Books

Patrick Moore, *Patrick Moore on Mars* (London: Cassell & Co., 1998).

> *Originally written in 1955 as Guide to Mars, and last revised just after the Viking landings, this book brings the story of Mars up to date. Moore covers the exploration of Mars from the time it was first observed through to the Viking missions, the results of which are examined in detail, and then turns to the newest findings, from the recent return to Mars. This book is an excellent introduction to Mars, both as a world and as an object seen through a telescope.*

More Advanced Reading

J. Kelly Beatty and Andrew Chaikin, eds., *The New Solar System*, 3rd ed. (Cambridge, Mass.: Sky Publishing, 1990).

> *A hugely informative book that will appeal both to the more advanced amateur astronomer and to college-level students. It contains three excellent essays about Mars that cover all aspects of its evolution, geology, and atmosphere: "Mars," by Michael H. Carr, "Surfaces of the Terrestrial Planets," by James W. Head III, and James B. Pollack's "Atmospheres of the Terrestrial Planets." In addition, various other articles treat Mars in the context of the other terrestrial planets. (Note that a fourth edition of this work appeared in 1998: see the references to chapter 5. The selection of essays varies somewhat across the two editions, however. I will cite the third edition when its contents seem especially pertinent to the subject at hand.)*

H. H. Kieffer, B. M. Jakosky, C. W. Snyder, and M. S. Matthews, eds., *Mars*, University of Arizona Press Space Science series (Tucson: University of Arizona Press, 1992).

> *The definitive view of Mars at the start of the 1990s, just before the recent wave of exploration. An advanced work for the real enthusiast that covers all aspects of the planet in depth.*

On the Internet

Mars
www.nineplanets.org/mars.html

> *A page about Mars from the Nine Planets Web site. A good place for the beginner in search of some basic information about the planet.*

The Mars Page of the National Space Science Data Center at Goddard Space Flight Center
http://nssdc.gsfc.nasa.gov/planetary/planets/marspage.html

> *This is the definitive Mars site, with links to space mission pages, images, and books that can be accessed online.*

Meteorites from Mars
www.jpl.nasa.gov/snc/

> *Extensive information on meteorites from Mars, with many links.*

The ALH84001 Meteorite
http://nssdc.gsfc.nasa.gov/planetary/marslife.html

Information, photographs, and links to stories about the famous ALH84001 meteorite that may contain evidence that life existed in the Martian past.

Meteorites from Mars
http://www-curator.jsc.nasa.gov/curator/antmet/marsmets/contents.htm

A site that compiles and analyzes scientific papers about ALH 84001. Although not recommended for the beginner, the site offers a lot of careful commentary on what the papers mean, making it a key resource for anyone who wishes to study the topic in detail.

The Martian Climate and Its Implications for Life
www.sigmaxi.org/amsci/articles/96articles/Zent.html

An article by Aaron P. Zent, originally published in American Scientist, *that examines climate change on Mars and the possibility of life. Zent proposes that Mars was once hospitable for life and explains how this could have come about. Written at an advanced popular level, the article contains a great deal of interesting information on Mars and its climate.*

CHAPTER 7

Pluto: Impostor or King of the Outer Darkness?

B ack in 1930, at the Lowell Observatory, in Flagstaff, Arizona, a young man from Kansas discovered a tiny speck of light in a photographic image of the constellation of Gemini. At the time, his discovery was hailed as the ultimate triumph of celestial mechanics. On the basis of its gravitational influence on the orbits of Uranus and Neptune, a ninth planet for the solar system had been found, just where it was expected to be. But not everyone was convinced by the new planet, which was soon named Pluto. Even the discoverer himself had doubts as to whether the object he had located was the planet he had expected to find. Since then, Pluto has continued to pose a problem for astronomers, generating a controversy that has at times reached the front pages of the newspapers.

In 1999, Pluto crossed Neptune's orbit, as it does every 248 years as it moves away from perihelion, and once again became the most distant planet in the solar system. This event served as a reminder of just how peculiar an object Pluto is—certainly no other planet in the solar system ever comes anywhere near crossing another's orbit—and led to renewed calls for Pluto's status as a planet to be reconsidered. Perhaps predictably, defenders of Pluto's current status interpreted this suggestion to mean that Pluto was to be downgraded from a major planet to a "small body." Even though nothing of the sort had been officially proposed by the IAU or one of its committees, the suggestion generated a heated debate. Most people forget that such a proposal is nothing new and that Pluto's identity has been questioned from the start. One of the main problems is that the dividing line between planets and small bodies (or minor bodies) is less sharp than we had thought and thus the definition of a planet far from clear. Various criteria have been proposed, but none has found universal favor—principally because they would either include, or exclude, Pluto. It seems astonishing that such a relatively tiny world can stir up such powerful feelings. At long last, in 2004, the IAU stepped in and formed a committee charged with formulating an acceptable definition of what constitutes a planet, although, as I write, we have yet to learn the result.

Many astronomers, myself included, feel a strong affection for Pluto. Small, remote, and mysterious, it inspires the imagination. Perhaps surprisingly, we know more about the major satellites of the planets than we do about Pluto. It is, at the start of the twenty-first century, the only planet not to have been explored and photographed by a space probe. Consequently, it is also the only planetary body of significant size not to have been properly mapped, in contrast to all the other planets and even most of their smaller satellites. At present it remains a question whether Pluto will be explored in the near future. Plans exist for a probe—the Pluto Express (now renamed "New Horizons"), designed at the Johns Hopkins University—that would be launched in 2006 and reach the planet in just nine years, as opposed to the fifty years that would otherwise be necessary, using Jupiter's gravity to speed it on its way. This mission has been very much a stop-and-go affair, however, and, after funding for the program was suddenly cut in October 2003 (having, only weeks before, been restored after a previous cut), it is by no means a foregone conclusion that the launch will go ahead. But if the probe cannot be launched in January 2006, it will miss the essential rendezvous with Jupiter, in which case it must launch directly to Pluto, taking an extra four years to arrive.

What is this strange little world? Is it a planet or an impostor? How did it come to be included among the major planets in the first place? Is it true that Pluto is, or was, in danger of being stripped of its planetary status? Are there more planets out there waiting to be found? Here we will try to discover some of the answers to these questions, although in many cases these answers will be no more than opinions. Please form your own, whether or not you agree with mine.

Out of the Darkness

How is it that the ninth planet came to be discovered by a Kansas farmhand? The answer shows how important luck and being in the right place at the right time can be, although it also shows that skill and dedication bring their own reward. Had the discovery been delayed for a few years, it might never have happened, or at least it might have had to wait several decades. As it was, the discovery of Pluto was in an important sense an accident—the culmination of a series of grave errors committed by astronomers over many years. At the time, though, nobody realized this.

To understand how Pluto came to be discovered one has to go back to the end of the eighteenth century.

The story of Pluto begins with a man born in Hanover, Germany. At the time, George II was king of England, but he also ruled over a significant area of modern-day Germany, including Hanover. Born in 1738, Friedrich Wilhelm Herschel belonged to a musical family. When he was just fourteen, he joined the band of the Hanoverian guards and, after his induction into the infantry, saw action in the Seven Years War, albeit only briefly. During this time, a posting to England as part of a force intended to guard against a French invasion allowed him to learn English and to forge a link with the country that would later adopt him as its own. In fact, people often assume that Herschel was English, and in fact he did eventually become a naturalized Briton, although not until the 1790s, as he was nearing the age of sixty.

Herschel returned to England in 1757, where he worked as a music copier and later as a musician, ending up in Bath. Here he rapidly developed a love of astronomy. On finding that a suitable telescope was well beyond his means, Herschel decided to build his own telescope, including the mirror. After some two hundred failures he finally produced a usable mirror, 13 centimeters in diameter, and in the years to come went on to build many more telescopes. Then, on March 13, 1781, a moment arrived that cemented his place in history and forever changed our solar system. Armed now with a 15.7-centimeter telescope—which would today be re-

Looking around Our Solar System

garded as very much a beginner's item—Herschel came across a star in the constellation of Gemini that seemed wrong. He described it thus: "While I was observing the small stars in the neighbourhood of H Geminorum, I perceived one that appeared visibly larger than all the rest; being struck with its uncommon magnitude, I compared it to H Geminorum and the small star in the quartile between Auriga and Gemini, and finding it so much larger than either of them, suspected it to be a comet."[1]

Herschel continued to observe his "comet" and established how its position changed from night to night. However, it quickly became obvious that it was not an ordinary comet. Despite Herschel's comment about its relatively large size, it was actually a tiny object that, for many other observers, working with inferior telescopes, appeared little different from a normal star. It moved extremely slowly from night to night, which suggested that it was a long way from the Sun. Within weeks, even before the official announcement of Herschel's discovery, a number of mathematicians had demonstrated that the new object was really a new planet, a finding that some astronomers had already anticipated on the basis of its movement, which was highly uncharacteristic of a comet. Herschel had discovered Uranus.

The impact of his discovery was fundamental. At a stroke, the size of the solar system had been doubled, as Uranus orbits the Sun at twice the distance of Saturn, the most distant planet known since ancient times. The existence of a new planet also dealt a devastating blow to astrology, which had only really begun to separate itself from the science of astronomy in the seventeenth century. One of the great bases of astrological mysticism had been the number 7: seven openings in the face (ears, eyes, mouth, and nostrils), seven "planets" in the sky (the Sun, Mercury, Venus, the Moon, Mars, Jupiter, and Saturn), and so on. With the discovery of an eighth planet, this symmetry was abruptly destroyed.

Strictly speaking, Herschel's discovery was an accident: he was neither searching for comets, nor was he searching for new planets. However, it was an *inevitable* accident. Herschel was reviewing the heavens, examining the whole sky with his telescope, on the lookout for any new or unusual object. Writing to a friend in 1781, he described the "accident" thus: "It has generally been supposed that this was a lucky accident which brought this star to my view; this is an evident mistake. In the regular manner I examined every star of the heavens, not only of this magnitude but many far inferior, it was that night *its turn* to be discovered. . . . Had business prevented me that evening, I must have found it the next." Only later was it discovered that Herschel was far from the first person to lay eyes on Uranus. Between 1690 and 1771 the planet had been observed more than twenty times by different astronomers in England, France, and Germany. This was wonderful news for astronomers because it meant that they now had observations spanning ninety-one years—more than the eighty-four years it takes Uranus to complete a full orbit. England's first astronomer royal, John Flamsteed, had observed Uranus no fewer than six times in 1690 and then again between 1712 and 1715, and the third astronomer royal, James Bradley, had recorded the planet in 1748, 1750, and 1753. Tobias Mayer saw it in 1756, and Pierre Lemonnier at least ten times between 1764 and 1771, including six observations in January 1769 alone. Some of these observations were, like Herschel's, of high quality, notably those of Flamsteed and Bradley, who made careful measurements and kept a thorough record of what they saw. In striking contrast, one of the observations made by the French astronomer Pierre Lemonnier was found scrawled on a paper bag that had been used for hair perfume.

A Comedy of Errors, I

As the earlier observations of Uranus were analyzed, however, astronomers began to realize that something was seriously wrong. As early as 1788, only seven years after Herschel's discovery, it had become clear that the orbit calculated for Uranus on the basis of these observations predicted a position that increasingly diverged from the planet's true position. One way to explain the discrepancy was to suppose that the older data were untrustworthy. As time wore on, and fresh observations accumulated, astronomers decided to throw away the old observations and use only those made since the planet was located. Initially, it seemed that things had improved, but as observations continued to be made, it again became obvious that Uranus was behaving badly.

What was happening? In 1834 an amateur astronomer in Kent, the Reverend T. J. Hussey, rector of Hayes, concluded that another, unknown planet was pulling Uranus out of position. In this he was absolutely correct. However, when he communicated this theory to the formidable George Biddell Airy, who within months would be appointed astronomer royal, the latter informed Hussey that it would be impossible to calculate the position of the unknown planet—and in this Airy

Looking around Our Solar System

was incorrect. Over the next few years Hussey's theory gained increasing support. However, nobody seemed to know how to tackle the problem of locating the missing planet.

Enter John Couch Adams. A mathematics student at Cambridge University, Adams became interested in the problem of Uranus in 1841 and decided to solve it. The word *brilliant* is sadly overused, but in the case of Adams it is more than justified. The Cambridge mathematics course is a searching test for anyone, and the title of "senior wrangler," bestowed on the person who finishes at the top in the final examinations, is highly prestigious. Adams's performance was astonishing. He was ahead by such a margin that, for the first time, the gap between the mark earned by the senior and the second wrangler was larger than that between the mark of the second and the last in the class.[2]

In February 1844, Adams requested, through one of his Cambridge professors, James Challis, the observations of Uranus that Airy had available. By September the calculations were finished, and Adams had solved the problem. Adams asked Challis to prepare him a letter of introduction so that he could visit Airy and present his work. Adams went to Greenwich three times, but he never saw Airy. The first time Adams arrived, Airy was returning from a meeting in Paris. The second time, Airy was in Greenwich, but he was not at home. All Adams could do was leave his card, which Airy's servant did not even pass along to him when he got home. Finally, Adams made a third trip to Greenwich. This time, Airy was at home, having dinner. But since he had no idea that Adams would be visiting, he had left no instructions that Adams should be allowed to see him, so Airy's servants simply told Adams that their master was not to be bothered. Adams left a paper describing his results and then essentially gave up. Airy subsequently read Adams's paper, with which he was evidently impressed, and responded with a minor question requesting clarification of a point that to Adams must have appeared trivial. This response served as confirmation to the diffident Adams that, despite his fine words of praise, Airy had little real interest in his work, and so Adams did not reply. And on not receiving a reply, Airy declined to pursue the matter.

A combination of bad timing, a servant's negligence, one man's shyness, and another's pride meant that Neptune was not discovered in England in 1844. Fortunately, the matter did not rest there.

The Rude Man

In 1845, the director of the Paris Observatory, François Arago, decided that it was time to resolve the problem of Uranus for once and for all and locate the object that was perturbing the planet. For this task he selected the formidable Urbain Jean Joseph Leverrier. Leverrier was brilliant, but he was also famed for his extraordinary rudeness and foul temper. Leverrier had earlier held the post of director of the Paris Observatory, a position of enormous prestige in the world of astronomy,[3] but he had been forced to resign because of his unacceptable behavior. In the opinion of a colleague, if Leverrier was not the most detestable man in France, he was certainly the most detested. His personality notwithstanding, Leverrier worked on the problem of Uranus with gusto and prepared three reports describing his results, which were published in December 1845 and in June and August 1846. Airy greeted the first report with fulsome praise, but at no point did he mention to Leverrier that Adams had already presented him with a solution more than a year earlier. On reading the second report, in which Leverrier specified a position for the planet almost identical to that Adams had worked out, Airy spoke in his private correspondence of his delight at seeing the calculations of Adams confirmed. He wrote to Leverrier, and he also discussed the problem with the director of Denmark's Seeberg Observatory, but to neither did he say anything about Adams's work.

In his third report, Leverrier offered his final and definitive solution. If Leverrier had expected his countrymen to rush to find the planet, he was to be disappointed. The prediction met with an underwhelming response. Unlike Adams, though, Leverrier was not to be deterred.

A Comedy of Errors, II

With two almost identical calculations of the position of the planet in hand, as well as a prediction that the planet would be both bright enough and large enough to show a noticeable disk, and with a major observatory at his command, one would have expected Airy to leap at the chance to observe that point of the sky and discover the planet.

He did not.

What followed was the most bewildering spectacle of incompetence and sloth.

For reasons known only to himself, Airy considered the telescopes at Greenwich, the largest of which was a 17-centimeter refractor, inadequate to the job of locating the planet—this for an object that was correctly predicted to be just a little fainter than Uranus and that is in fact easy to see with small binoculars. So Airy decided to enlist the services of the Cambridge Observatory, with its famous 30-centimeter Northumberland reflector. The man charged with the search was none other than Adams's sponsor, Challis, a man who should have had an enormous vested interest in success. Airy wrote to him twice in July of 1846, the second time with great urgency. Challis was away when both letters arrived, but they eventually reached him. At this point, one would assume that Challis—faced with two almost identical predictions—would immediately check the appropriate point in the sky to see whether there was anything unusual.

He did not.

Instead, Challis decided to search a large area of sky around the predicted position. But as he had no suitable star map of this area, he resolved to prepare one himself. This he did in a slow, methodical, and rather lethargic way, admitting that he would certainly not be able to complete his search in that same year. Even more unforgivable, he did not check his observations against each other to see whether any "star" shifted its position from night to night. Challis observed Neptune twice in his first four nights of searching in early August 1846, as well as a third time later on, and on all three occasions he managed not to recognize it. Had he only compared his observations, however, searching for any stars that had moved, it was there to see—and on eventually hearing of the discovery he quickly located the planet on his maps.

One rather curious incident, which seems to be relatively little known, amply illustrates how little importance Challis evidently attached to his search. One night he noticed a "star" that appeared to show a small disk, but instead of checking on it he simply noted the fact, with the avowed aim of checking it "tomorrow." The following night he discussed the observation over dinner with the chaplain of Sidney Sussex College, the Reverend William Kingsley. Kingsley made the obvious suggestion that Challis should look at the star again with a more powerful eyepiece. Challis agreed, although quite why the voice of the church was needed in order to make this action to seem logical is uncertain. The evening was clear, and the great discovery awaited. Challis's wife pressed further after-dinner cups of tea on them, and the conversation flourished. By the time the teapot had finally been emptied, the sky had clouded over. As an Englishman who likes his tea, I under-stand what it means to be "dying for a cup of tea"—but this was ridiculous.

If that were not enough, when Challis did finally check on the star, he dis-covered it was no longer there. Of course, a moving star that shows a disk could only be a planet, but even in the face of this evidence Challis failed to act. It was, of course, Neptune.

Leverrier presented his final results to the French academy on August 31, 1846. The contrast with Airy, Challis, and Adams could not be starker. When a search was not instituted immediately, Leverrier wrote to Johann Galle at the Berlin Observatory with an appeal for assistance. The letter arrived on September 23. That same night Galle pointed the telescope at the point predicted by Leverrier. While Galle described the stars he could see, a young student called Heinrich D'Arrest, who had asked to assist him, compared Galle's descriptions with the map of the sky. Within minutes he famously exclaimed, "That star is not on the map!"

Neptune had been found. It was half a degree from Adams's predicted posi-tion and less than a degree from Leverrier's, and it was almost exactly of the di-ameter predicted by Leverrier. As the discovery of Pluto would later be as well, this result was heralded as a triumph for celestial mechanics. For the first time an unknown body had been located mathematically, by means of calculations based on the influence of its force of gravity. Or had it? We now know that the answer is not as straightforward as it originally seemed. At the time, however, the solar system appeared complete, and the movement of Uranus was at last fully explained.

The Unanswered Questions

The story of the discovery of Neptune is riddled with unanswered questions. Many center on the role of George Biddell Airy. Why did he apparently take Adams's work so lightly? Especially given the long-standing Anglo-French rivalry, which had recently led to several wars, when Airy received an almost identical pre-diction from Leverrier, why did he not order an immediate search of that point in the sky? Why did Airy later praise the work of Leverrier but ignore that of Adams, which he had in fact received earlier? Why did he not mention Adams's work to Leverrier, if only to confirm to him the reliability of his conclusions? Why, when the predictions were that the missing planet would be almost as bright as Uranus and thus visible with the smallest telescope, did Airy consider that a large telescope would be needed to see it? Some have defended Airy on the grounds that at the time he was heavily occupied with public duties other than astronomy, most particularly as a member of the Railway Gauge Commission—but, even so, some of his actions were at best peculiar.

Of course, to be fair, other participants in the affair did not cover themselves in glory either. Adams was curiously diffident and lacking in persistence. When Airy let him down, he could have asked Challis to search himself. And then, on seeing Adams's results, why didn't Challis quickly embark on a search in order to gain himself and his observatory the prestige of discovering a new planet? Moreover, when Airy finally asked him to search for the planet, Challis was notably lackadaisical in his efforts and, one could argue, showed himself thoroughly incompetent, despite having been Adams's sponsor with Airy.

There is another mystery, however, that is particularly hard to understand. At the start of the century, six amateur astronomers—five Germans and the Hungarian Baron Franz Xaver von Zach—were called by Johann Schröter to a famous meeting at his observatory in Lilienthal to discuss the problem of the missing planet between Mars and Jupiter and decided to organize a search for the planet. Between 1801 and 1807 the group, who came to be known as the Celestial Police, located three of the first four asteroids ever to be discovered in the course of deliberately scanning the sky looking for new planets.[4] By the time Neptune was finally recognized, on September 23, 1846, a fifth asteroid had been added, and the number was doubled within two years. But why did the Celestial Police not find Neptune, which was much brighter than several of these asteroids?

We will never know the answers. But what is certain is that Neptune should have been discovered long before it was.

Pickering's Planets

With the discovery of Neptune all seemed well with the orbit of Uranus, at least for a time. But, as had earlier been the case with Uranus, astronomers soon realized that Neptune had been observed on a number of occasions before it was officially discovered. Apart from Challis's observations, the French astronomer Joseph-Jérôme de Lalande had observed Neptune on May 8 and 10, 1795, and had even noticed a discrepancy in his observations: one star (Neptune) had shifted its position between one night and another, although Lalande presumably put this down to an error in measurement. John Lamont, a Scottish astronomer, saw the planet on October 25, 1845, and then again on September 7 and 11, 1846—just days before the discovery by Galle and D'Arrest.[5] On the basis of these observations astronomers were able to define Neptune's orbit more precisely and thus its perturbations of Uranus's orbit as well. When these perturbations were taken into account, it initially appeared that Uranus followed its predicted path, but slowly the discrepancies crept back. Gradually the conviction grew that there must be yet another planet waiting to be discovered.

Indeed, this possibility had already been raised as early as 1834, twelve years before Neptune was discovered. By the 1870s many astronomers were convinced that there was at least one other planet out there, and a number of searches had been undertaken, although to no avail. Around the start of the twentieth century, the problem was taken up by two Americans, at one point friends and collaborators but later bitter rivals: Percival Lowell and William Pickering. Both had some unusual ideas and are often best remembered for their mistakes. Lowell's views on the Martian canals severely damaged both his reputation and that of the Lowell Observatory, while Pickering was convinced that there was a detectable lunar atmosphere. Both studied the problem posed by the orbits of Uranus and Neptune and made several predictions about the position of possible additional planets. Pickering made his first such prediction—Planet P—in 1911 and latter added a second planet, Planet O. Lowell published his own report on Planet X in 1915.

Much has been made of the fact that the orbit of both Lowell's Planet X and Pickering's Planet O bore a striking resemblance to the peculiar and highly unexpected orbit of Pluto. In fact, when Pluto was finally discovered, a symbol was adopted for it—superimposed P and L—that stood not only for the first two letters of the planet's name but also for the names of Pickering and Lowell. However, when their predictions were examined more closely, it became evident that both men were on shaky mathematical ground. In particular, Pickering's calculations were of dubious validity.

Pickering noted that certain comets had a variable orbital period. We have already seen how, despite the best efforts of those attempting to calculate its orbit, predictions for the return of Halley's comet in 1910 showed an inexplicable error of three days. Similar inconsistencies were noted with other comets, leading astronomers to suggest that an unknown planet was pulling them out of position in the same way that Neptune pulled Uranus out of position. Pickering made a whole series of predictions regarding this mystery planet, although these tended to change both drastically and with some frequency. His two most important candidates were designated Planet P and Planet O. Planet P was presumed to be a large planet at a great distance from the Sun, while Planet O was a much smaller body close to the orbit of Neptune. Pickering initially estimated that Planet P should be approximately 123 AU from the Sun, four times the distance of Neptune, but his estimates ranged from there down to 75.5 AU. (An AU, or astronomical unit, is equivalent to roughly 150 million kilometers.) When he issued his final prediction concerning Planet P, in 1931, he assigned it a mass three times greater than that of Neptune, or fifty times that of Earth. The planet would have had a magnitude of about 11 and should therefore have been easy to detect.

Planet O was a different kettle of fish. Pickering predicted an eccentric orbit that would cross Neptune's, a mass twice that of the Earth, and a magnitude around 15. In 1919 Pickering persuaded the director of Mount Wilson Observatory to search the relevant area of the sky with the observatory's 25-centimeter telescope. Although only four photographic plates were taken, Pluto was in fact visible in all of them. However, the images were not inspected as thoroughly as would have been necessary to locate the four faint images of the planet. In the meanwhile, though, Pickering substantially revised his prediction. According to his new theory, the planet was smaller than the Earth, not twice its size, and it was also somewhat closer to the Sun than he had originally envisaged, such that it would spend half its orbit closer than Neptune to the Sun. It would therefore be considerably brighter than originally predicted.

What Pickering did not know, and could not have known, was that the perturbations of the orbit of Halley's comet and others, such as Comet Westphal and Comet Pons-Gambert, had nothing to do with any unknown planet. As we saw in chapter 4, the explanation was the tiny rocket forces produced by gases emitted from the comet's nucleus. Pickering's planets did not exist. Moreover, nothing could hide the fact that his calculations had changed so often and so much that by 1930 nobody was willing to take them seriously. In the end, this lack of credibility was to halt all his efforts to get a search for his planet instituted.

And what of his now bitter rival, who was also suffering a severe lack of credibility after the fiasco of the Martian canals?

Lowell's Planet X

Lowell adopted a completely different method. He had initially attempted to calculate the position of the missing planet on the basis of anomalies in the orbits of comets but subsequently discarded that approach. Having noted that there was still a small discrepancy in the position of Uranus even after the gravitational pull of Neptune was factored in, he used this discrepancy as the basis for his search for a new planet. Lowell was, for all his faults, systematic. In constructing the observatory that bears his name, for example, he began by looking for the best possible site at high altitude, an approach that was many decades ahead of its time. Lowell set out to calculate the position of the new planet by taking into account all the observations of Uranus available to him. In the days of hand calculation this was a gargantuan task.

In 1914 he published his results, in which he proposed the existence of what he called Planet X. The position he first gave for the planet was in the constellation of Libra, deep in the Milky Way. A number of photographic images were taken at Lowell Observatory in an effort to detect the planet, but without success. A year later, in 1915, Lowell suggested an alternative position in Taurus, close to its border with Gemini, separated by 180 degrees in the sky from the one initially proposed. Once again, Planet X would be located in an area crossed by the Milky Way and would thus be well hidden. The orbit Lowell predicted for his planet in 1915 was actually remarkably close to the orbit of Pluto. Between 1914 and 1916 almost a thousand photographic plates were taken in an attempt to find the planet—a huge undertaking—but nothing was found.

Percival Lowell died of a stroke in 1916 at the rather young age of sixty-one. He was a tired and discouraged man. Had he known that Pluto was faintly, but clearly, visible in two of the nearly one thousand photographs taken at his observatory, on March 19 and April 15, 1915, his story would have had a different ending. Indeed, this is one of the wonderful "what ifs" of astronomy. What if Pluto *had* been recognized on the plates taken by Lowell's collaborators in 1915—or on those taken for Pickering at Mount Wilson in 1919? As it was, however, with Pickering largely discredited, Lowell dead, and all the searches that had been made unsuccessful, it now seemed that it might be many years before the missing planet was found.

Looking around Our Solar System

The Kansas Farmhand

After Lowell's death, his observatory passed through some desperate moments. Because of the Martian canals episode, other astronomers did not for the most part take the work of the Lowell Observatory very seriously. Understandably, the Lowell staff felt isolated and unhappy at finding themselves largely ignored by the astronomical community. Lowell left three dedicated assistants to carry on his work, all of whom are remembered today as astronomers of great ability. In particular, Vesto Slipher, the new director of the observatory, made important contributions to astronomy with the telescopes in Flagstaff. On his death, Lowell left the observatory an endowment of $1 million, which proved vital in what was to follow, and Lowell's family continued as trustees. Sadly, despite being generously provided for, Lowell's widow challenged the will, and much of the endowment was squandered in legal fees, to the extent that the observatory lacked the funds for a new telescope—a needed item if the search for the missing planet was to continue. Finally, in 1929, a gift of $10,000 from the late Percival's brother, Abbott Lawrence Lowell, president of Harvard, allowed the telescope to be constructed and begin operation.[6] The new telescope was a fine 33-centimeter photographic refractor capable of producing high-quality images of a $14° \times 12°$ area of sky on large glass photographic plates.

There was only one question. Who would have the time to operate this telescope and spend night after night taking one-hour exposures all around the sky? None of the observatory staff could spare much time from their other research. The answer arrived from a most unexpected source. In Kansas, a young man of twenty-two, Clyde Tombaugh, was pondering his future. Fascinated by astronomy, he had built his own telescope. Unfortunately, in the summer of 1928 a crop disaster caused by a severe storm had left his parents struggling for money. Tombaugh hired himself out to a neighbor as a farmhand, but at the same time he was looking for a way out of farming. He had read the reports on Mars written by the Lowell staff, particularly Vesto Slipher's brother, Earl Carl Slipher, in a back issue of *Popular Astronomy*. That autumn he observed Jupiter and Mars, and he sent his drawings to Vesto Slipher. Slipher was evidently impressed and immediately decided that this was the person to operate the new telescope that would soon be arriving. He wrote back to Tombaugh with a series of questions. When Tombaugh's replies proved satisfactory, a second letter arrived that asked, surprisingly, "Are you in good health?" Finally, Slipher sent a third letter inviting Tombaugh to come to Flagstaff in January for a trial period of a few months as the operator of the new telescope.

Amazed by his luck, Tombaugh jumped at the chance and on January 14, 1929, started the lonely 1,600-kilometer journey by train to Arizona. Slipher himself met the nervous youngster, who had never previously been away from home, and took him to the observatory where, apart from learning to operate the telescope, he was expected to do odd jobs such as stoking the furnace and shoveling snow. On April 6, 1929, Tombaugh used the new telescope to take the first photographic plates of the area of the sky in which the missing planet was presumed to be. On April 11, on just the tenth plate to be exposed, he in fact had his first image of Pluto, albeit unbeknownst to him. On April 30, he photographed the region for a second time so that the two images could be compared, and again he captured Pluto. Unfortunately, the extreme cold cracked the first plate, and it could be used only with extreme difficulty. The plates were examined rapidly by the two Sliphers, but they missed the planet among the tens of thousands of stars on the images. Once again, Pluto escaped unseen. For the third time we can wonder, if only . . . For the Lowell team, the discovery of Planet X was turning into the holy grail, the object that would regain the honor and status of the observatory in the eyes of the world. It was a nerve-wracking period.

As the number of plates mounted, Tombaugh thought to himself that someone would have a huge job to do in checking them all. Prophetic words. No prizes for guessing who—what with the summer arriving and no plates analyzed—was given the job. Like Challis before him, Tombaugh was methodical. Unlike Challis, he was incredibly industrious. There is no question that Pluto was found thanks to his capacity to work out the best and most efficient way to do everything connected with the project.

Finding the Tiny Dot

To aid the search for Planet X the observatory had a blink comparator, a simple but ingenious device that enables two photographic plates of the same area to be compared. A small motor moves a mirror, allowing the observer's field of vision to flip between the two. Any object that moved between the two exposures will jump, while the stars remain still. Tombaugh decided that an interval of about

Looking around Our Solar System

six days between exposures would cause the planet to jump a sufficient distance to be detected comfortably by the operator viewing the two exposures. In their spare time, Slipher and his brother shared the job of blinking the first few frames in the area around Lowell's estimated position for the planet. Then, on June 18, Slipher asked Tombaugh to start the job in earnest. A week or so later Slipher and Tombaugh were showing a visiting astronomer the new telescope and explaining about the search for the planet. At one point Slipher had to go outside to answer the telephone. The astronomer leaned to Tombaugh and stated confidentially: "Young man, I am afraid that you are wasting your time. If there were any more planets to be found they would have been found long before this."[7]

Tombaugh's spirits were deflated, but he responded stubbornly that he was going to give the search his best efforts. As the year progressed, the region being photographed moved around the sky, tracking the opposition point—the point in the sky directly opposite the Sun from the Earth. In January 1930 Lowell's target region in Gemini came round again. The first attempt to photograph it, on January 21, was spoiled by poor conditions, but new frames taken on January 23 and 29 were perfect. On February 18, Tombaugh decided to examine these plates, which centered on the star Delta Geminorum, mainly because the other photographs of this region showed so many stars that he could not face the prospect of looking at them all. From such trivial details do great discoveries come. At 4:00 P.M., give or take a minute or so, Tombaugh noticed a tiny star of magnitude 15 that jumped as the machine blinked three times per second between the January 23 and 29 images—except that stars don't jump. It was Pluto. After checking the images with great care, Tombaugh walked calmly to Slipher's office and announced: "Dr. Slipher. I have found your Planet X. I'll show you the evidence."[8] As Tombaugh himself later remarked, had he not made the rather arbitrary decision to examine the area around Delta Geminorum, it is unlikely that Pluto would have been discovered in 1930—although, with the pressure to discover the planet mounting, it is hard to know what would have happened.

But was this really Planet X? Although this is often forgotten today, even the Lowell team had their doubts.

To Be or Not to Be Planet X

Tombaugh knew immediately that his "star" was more distant than Neptune, simply by the slowness of its movement, but it seemed worryingly small and faint to be Planet X. Neither the discovery telegram, sent to the IAU on March 13, 1930, nor the hastily prepared observations circular providing full details of the discovery that was sent to hundreds of universities and observatories used the word *planet* to describe the object. It appears that the first comment to the effect that the object was a new planet was made, not by astronomers, but in the Associated Press wire that communicated the story. Although there is little doubt the Lowell staff were confident that the new object was a planet, they wished to measure its size and calculate its orbit before making such a claim.

After four days of frantic work by the three senior members of the Lowell staff and one of their former instructors, from a course on planetary orbits they had taken as students many years before, Pluto's orbit had been calculated. Their first calculation, which they based on data from just a few observations, produced some dismay. Pluto's orbit appeared to resemble a comet's more than a planet's, which made the Lowell team fear that the object might not be a planet after all. Lowell and Pickering had also predicted that the missing planet would have a mass one to several times that of Earth, but on an initial estimate the mass of this object was no more than one tenth of Earth's mass, comparable to Mars, and far too small to perturb Uranus and Neptune.

Both Harlow Shapley at Harvard and A. O. Leuschner at the University of California at Berkeley suggested that Pluto might be the first of a new class of what later became known as trans-Neptunian objects (TNOs), namely, small objects that would orbit the Sun beyond the orbit of Neptune—an inspired suggestion, we now know, that took over sixty years to prove. (Interestingly, one of Leuschner's assistants in calculating the orbit of Pluto was a young student called Fred Whipple, later to become famed for his work on comets.) In the meanwhile, the Lowell staff were now seriously worried about the possibility that Pluto might be a distant comet and examined it carefully with the 61-centimeter telescope to see whether they could detect a fuzzy coma around the image. What killed the giant comet theory, though, was Pluto's orbit. Once again, in the weeks following the announcement of the discovery, astronomers at other observatories realized that they had photographed Pluto in the past without being aware of it. In fact,

Table 7.1   Pickering's Planet O, Lowell's Planet X, and Pluto

	Pickering[a]	Lowell	Pluto
Mean distance from Sun	55.1 AU[b]	43.0 AU	39.5 AU
Orbital eccentricity[c]	0.31	0.20	0.25
Inclination	15°	10°	17°
Longitude of the ascending node[d]	100°	—	109°
Longitude of perihelion[d]	280.1°	204.9°	223.4°
Mean annual motion[e]	0.880°	1.241°	1.451°
Period	409 years	282 years	248 years
Perihelion date	2129.1	1991.2	1989.8
Longitude epoch 1930.0[f]	102.6°	102.7°	108.5°
Mass (Earth = 1)	2.0	6.6	0.0025[g]
Magnitude	15	12 to 13	15

SOURCE: Adapted from William Graves Hoyt, *Planet X and Pluto* (Tucson: University of Arizona Press, 1980), p. 222.

[a] Both Pickering and Lowell were prone to revising their predictions concerning the orbit of their respective planets. In such cases, the prediction that came closest to the actual orbit of Pluto has been preferred.

[b] An AU, or astronomical unit, is approximately 150,000,000 kilometers (more precisely, 149,597,900).

[c] Orbital eccentricity is defined as a relationship between the smallest distance and the average distance of an object from the Sun.

[d] The ascending node is the point in an object's orbit where it crosses the plane of the Earth's orbit—the ecliptic—as it moves from the south to the north; perihelion is the point in its orbit where it is nearest the Sun. Both the longitude of the ascending node and the longitude of perihelion are measured as angles from the "Greenwich meridian" of the solar system, namely, the line from the Sun to the Earth at the spring equinox. To determine the two longitudes, one draws a line from the Sun either to the position of the ascending node or, as the case may be, to the point of perihelion and then measures the angle between that line and the meridian. Together, the two angles are used to define the orientation of an object's orbit.

[e] An object's mean annual motion is the number of degrees it moves each year as it orbits around the Sun. If the orbital period were 360 years, the mean annual motion would be exactly one degree. Thus, an object's mean annual motion will be less than one degree if an object's period is longer than 360 years and more than one degree if its period is shorter than 360 years.

[f] The longitude epoch is an object's longitude in its orbit at a particular point in time, in this case 1930.0, or the beginning of 1930. (The convention in astronomy is to express dates as a year followed by a decimal fraction—that is, to divide the year into tenths rather than into the usual twelve months.) Interestingly, the positions that Pickering and Lowell predicted for their planets in 1930 are in remarkably close agreement (102.6° and 102.7°) and are also not far different from Pluto's actual position (108.5°). This appears to have been a coincidence, given that the calculations Pickering and Lowell used to arrive at these numbers were completely off base. Not surprisingly, then, when it comes to the other parameters, large discrepancies often exist between their respective predictions and between those predictions and the figures for Pluto itself.

[g] Note that Pickering's estimate of Pluto's mass was 800 times too large. But he was still closer than Lowell, who was out by an astonishing factor of roughly 2,500.

within a few weeks images dating back as far as 1908 had been unearthed from observatory archives. With the help of these observations, a much more accurate orbit was calculated for Pluto—practically identical to the one accepted today. Although the orbit crossed Neptune's and was rather eccentric and inclined, it was clearly closer to the orbit of a planet than a comet. The original, very eccentric, cometlike orbit was merely the result of trying to fix the path of such a distant body using too few observations.

Grave doubts remained, however, both at Lowell and elsewhere, about the new object. Might Pluto be the satellite of a much larger body? Could there be other planets waiting to be found? After a brief hiatus the staff at the Lowell Observatory decided to resume the search for Planet X. First Tombaugh searched the surrounding regions to make sure that the "real" planet was not lurking undiscovered, as would have been the case if Pluto were merely a distant satellite of another planet. Tombaugh was able to show that there was no other body, brighter or fainter than Pluto, within 5° in the sky. Whatever it was, then, Pluto was not a satellite of a larger body. A little later, having discounted this alarming possibility, Slipher asked Tombaugh to continue searching in case more bodies were there to be found in the trans-Neptunian region—denizens of what is often called the "Kuiper Belt," for the Dutch astronomer, Gerard Kuiper, who first suggested that debris similar to that found in the asteroid belt might exist in the distant region beyond Neptune. Tombaugh went on with his work until 1943, in which time he covered two-thirds of the entire sky, most of it to a limiting magnitude of about 17. At that point, he had searched all of the sky to 30° north and south of the ecliptic plane (the plane in which the planets orbit) and a substantial fraction of the sky further north and south, omitting only the northern and southern polar regions. During this extended search he discovered, in the order of importance that he later assigned, a globular star cluster, a supercluster of galaxies, several small clusters of galaxies, five galactic star clusters, a comet, and about 775 asteroids. But he did not find any more trans-Neptunian objects. This last, however, was due to an odd circumstance—a cosmic conspiracy of sorts.

By 1943 Clyde Tombaugh was convinced that there were no more trans-Neptunian bodies of significant size. He believed, correctly, that there were no more objects even a tenth as bright as Pluto to be found. However, he also thought, incorrectly, that this was because there were no more large bodies to be found. Of course, we now know that there are millions of objects beyond Neptune. In fact, as of February 27, 2005, there are, including Pluto, 945 TNOs. Of these, two—designated 2002 LM_{60} and (28978) Ixion—appear to be more than 1,000 kilometers across, a diameter that has been suggested as one possible dividing line between a genuine planet and a smaller body. So far none rivals the 2,274-kilometer diameter of Pluto, although 2002 LM_{60} and (28978) Ixion have a diameter approximately half the size.[9]

But if these objects are so large, Tombaugh should have detected them easily, especially on his deep search. So why didn't he? The reason is that nature played a trick on him. Pluto is a highly reflective object. Its surface is covered with methane frost that is brilliantly white. But the other TNOs are dark, and it is assumed that most of them are as black as coal. Pluto reflects at least ten times as much light as they do and so appears much brighter, although it is only a little larger than other TNOs. In short, Tombaugh saw Pluto because it stands out. Were Pluto as dark as a typical TNO it might very well have been too faint for him to detect. It is in fact easy to show that a dark object as large as Pluto might well escape detection if it were just a little further away from the Sun. Similarly, had Pluto been at the most distant point of its orbit instead of being close to perihelion it would also have been difficult for Tombaugh to detect.

In the ongoing debate about whether Pluto is genuinely a planet, some argue that it would be a slight to Clyde Tombaugh's memory if Pluto were demoted from planetary status. Indeed, it is instructive to compare the search made by Challis for Neptune with that made by Tombaugh for Planet X. Challis checked 3,100 stars and saw Neptune three times without recognizing it. Tombaugh checked 45,000,000 stars and recognized Pluto the first time he saw it. There is, as they say, no contest. Whatever history's verdict is about Pluto, Tombaugh's feat is without doubt the greatest in the history of observational astronomy. On the single photographic plate alone where Pluto was located there were a hundred times more stars than Challis ever had to check.[10]

A Mathematical Accident

Initially, the discovery of first Neptune and then Pluto according to predictions based on perturbations of the orbits of known planets appeared to be a striking confirmation of the predictive power of mathematics. In the case of Pluto, however, we now know that the agreement of the prediction with the location of the planet was a coincidence. The planet Lowell had envisaged was about 2,500 times more massive than Pluto. As it is, Pluto could not conceivably perturb the motion of Uranus and Neptune to any measurable degree. Tombaugh's discovery was the result of a brilliant effort and a just reward for extraordinary persistence and dedication. But Pluto, as we have seen, was not Lowell's Planet X, and in that sense its discovery was an accident. But what about Neptune?

The perturbations of Uranus by Neptune were real, and the deviations of Uranus from its orbit correctly measured. That much is not in doubt. What is not often noted, though, is that both Adams and Leverrier used an erroneous method to calculate the position of the planet. Both assumed that its distance from the Sun would accord with the so-called Titius-Bode relationship, sometimes referred to as Bode's Law, which describes the regular progression in their distance from the Sun of the planets Mercury through to Uranus. But this law does not work for either Neptune or Pluto. If one follows Bode's Law to calculate the distance of Neptune from the Sun and uses that as a starting point, one may be able correctly to predict its position at one date, but the prediction will be utterly wrong for other dates.

This is, in fact, precisely what happened. Adams made no fewer than six different calculations, three of which were hopelessly in error in predicting the position of Neptune in 1845. For dates prior to 1810 and after 1875 all six calculations were out by more than 5 degrees, and some were *far* worse than this, with errors in the tens of degrees. Similarly, Leverrier's calculations are accurate around 1845, but if one goes back to 1800 they are seriously in error—by almost 25 degrees. Leverrier calculated that Neptune's orbit would be eccentric, which it is not, and that its orbital period would fall between 207 and 233 years, whereas it is actually 168 years, a difference four times Leverrier's calculated margin of error. Rather than a triumph for celestial mechanics, the discovery of Neptune, which inspired the search for Planet X, was actually a rather embarrassing case of luck. Had astronomers realized this at the time, it is probable that the discovery of Pluto would have been delayed by at least fifty years, until modern computers and electronic observing techniques would finally have made it inevitable.

How Do We Know about Pluto?

Nobody has seen Pluto close up. President Nixon canceled the proposed "Grand Tour" (which later became the Voyager missions) that would have seen one mission go to Jupiter, Saturn, and Pluto, a second travel to Jupiter, Saturn, Uranus, and Neptune.[11] As a result, no spacecraft has ever visited Pluto. All the same, we have been able to learn a considerable amount about this distant world, and in this luck has again played its part.

In 1978 James Christy, who was working at the U.S. Naval Observatory's Flagstaff station, made a startling discovery. He had taken a series of high-resolution photographs because he had to measure the position of Pluto against nearby stars with the greatest possible precision. His aim was to identify stars that might be briefly occulted by Pluto as it moved round its orbit. which would allow the size of Pluto to be measured on the basis of how long the star remained hidden. For the purpose of making these measurements, photographs taken in April and May were greatly enlarged and then examined closely. One can imagine Christy's dismay when, instead of seeing nice, round images of Pluto, he discovered that many of them were pear-shaped and thus useless for taking accurate measurements.

Initially Christy thought that there had been a problem with the exposure— that the telescope had not tracked Pluto's movement correctly. The bulge in the images was sometimes in the wrong direction to be the result of bad tracking, however, and the surrounding star images *were* perfect. Furthermore, the bulge appeared in only some of the images, and it was not always in the same direction. Christy started to suspect that what he was seeing was evidence that Pluto had a satellite, probably quite a good-sized one. He therefore checked photographs from 1965, 1970, and 1971, which all showed the strange, moving bulge when examined under magnification. Although a few astronomers still had their doubts, most accepted the news that Pluto had a large satellite that orbited the planet once every 6.39 days—the equivalent of Pluto's rotation period.

Pluto's satellite was named Charon, after the ferryman who carried the dead across the River Styx to the underworld, which was governed by Pluto. (The name is pronounced "Sharon," however, for Christy's wife.) It is very close to Pluto and, except with the benefit of special techniques, it can be detected only every three days, when at its greatest distance from Pluto, first on one side and then on the other. Even so, for many years some astronomers continued to doubt its existence, until the Hubble Space Telescope finally photographed the two as separate objects.

Once a satellite had been discovered, its orbit could be used to measure the mass of Pluto and Charon simply by following the laws of gravitation, which establish a relationship between how widely separated two bodies are and how quickly they orbit around their common center of gravity. The more massive the two bodies are, the stronger the pull they exert on each other, and so the faster they must move in order to maintain a stable orbit. Thus, for any particular separation, the greater the speed at which the two bodies are moving, the greater their combined mass must be. When these calculations were carried out for Pluto and Charon, the result came as quite a shock. Together, Pluto and Charon have just one-fifth of the mass of the Moon—only a tiny fraction of even the most pessimistic previous estimate of Pluto's mass. Whatever was perturbing Uranus and Neptune, it was not Pluto.

What about Pluto's diameter? On June 9, 1988, Pluto passed in front of a star in Virgo with a magnitude of 12. Many telescopes in Australia and New Zealand and also NASA's flying observatory—the Kuiper Airborne Observatory—were able to observe the occultation. The observations showed the presence of a tenuous methane atmosphere, about one-millionth of the density of the Earth's atmosphere at sea level, which dimmed the star gradually for about fifteen seconds before it disappeared behind the planet. This atmosphere is caused by the methane ice on the surface of Pluto, which sublimes—changes directly from a frozen solid to gas without ever becoming liquid—when the planet is close to perihelion, as it was 1988. For most of its orbit, though, Pluto has no detectable atmosphere. Since we know precisely what the velocity of Pluto is in its orbit, it is a simple exercise to calculate how big Pluto is from the length of time the star was hidden. The planet turns out to have an estimated diameter of 2,230 kilometers, with an error of only about 8 percent. This means that Pluto is less than half the diameter of Mercury and a lot smaller than the Moon. In fact, as Table 7.2 shows, no fewer than seven planetary satellites are larger than Pluto.

Is Pluto a Planet?

In 1999, a major international hue and cry arose in astronomy circles when a proposal was made to mark the occasion of the official numbering of the ten-

Looking around Our Solar System

thousandth asteroid by giving this landmark number to Pluto—a proposal that was misinterpreted to mean that Pluto was being downgraded from a planet to an asteroid. In fact, the intention was to recognize Pluto's unique status as both a planet—the King of the Outer Darkness—and a trans-Neptunian object, the latter by giving it the honorific number of asteroid 10,000. In view of the fact that TNOs had been steadily increasing in number, it was clearly time to begin including them among the asteroids, even though, in contrast to the rocky objects that circle between Mars and Jupiter, they are presumably icy bodies (as are Pluto and many of the satellites of the outer planets). In the future, then, the best-observed TNOs—the ones that had been observed often enough and whose orbits had been fairly accurately established—would be assigned specific numbers in the same way as ordinary asteroids. At the time the proposal was made, the number of asteroids had reached 9,999, and the number remained there while the controversy over Pluto raged.

The IAU's Minor Planet Center held an unofficial Internet vote on the issue, which resulted in a 63 percent to 37 percent split in favor of Pluto becoming asteroid number 10,000. Interestingly, opinion was sharply divided along national lines, at least as far as it was possible to identify the country of origin of the votes. In the United States—which contributed a large fraction of the total—the margin was very narrow: only 51 percent to 49 percent in favor of the proposal. Astronomers in Europe, however, tended to be far more sympathetic to the idea. Writing in the *New York Times*, Malcolm Browne made the following wry comment about the phenomenon: "If Pluto had been discovered by a Spaniard or Austrian, I doubt whether American astronomers would object to reclassifying it as a minor planet. Before he died, Clyde Tombaugh himself said he was reconciled to the perception of Pluto as one of many Kuiper Belt objects—minor denizens of the outer solar system."[12]

As we have seen, Pluto is small even in comparison to Mercury or to the largest satellites of the solar system. In fact, as Table 7.2 shows, no fewer than seven planetary satellites have a diameter greater than Pluto's. Had astronomers known this in 1930, what with the doubts that already existed about Pluto's status, it is certain that Pluto would never have been classified as a major planet. In fact, the main reason for classifying Pluto as a major planet is historical ("It will confuse people if we change its status"). Such a change in status would not be completely unprecedented, however, for Ceres was briefly regarded as a planet in the early nine-

Table 7.2   Small bodies of the solar system

Object	Type	Diameter[a]
Ganymede	satellite of Jupiter	5,262 km
Titan	satellite of Saturn	5,150 km
Mercury	planet	4,878 km
Callisto	satellite of Jupiter	4,800 km
Io	satellite of Jupiter	3,630 km
Moon	satellite of Earth	3,476 km
Europa	satellite of Jupiter	3,138 km
Triton	satellite of Neptune	2,700 km
Pluto	unclear	2,300 km[b]
Titania	satellite of Uranus	1,580 km
Rhea	satellite of Saturn	1,530 km
Oberon	satellite of Uranus	1,520 km
Iapetus	satellite of Saturn	1,440 km
2002 LM$_{60}$	trans-Neptunian object	1,200 km
Charon	satellite of Pluto	1,190 km
Umbriel	satellite of Uranus	1,170 km
Ariel	satellite of Uranus	1,160 km
Dione	satellite of Saturn	1,120 km
(28978) Ixion	trans-Neptunian object	1,065 km
Tethys	satellite of Saturn	1,050 km
(1) Ceres	asteroid	914 km
(20000) Varuna	trans-Neptunian object	900 km
2002 AW$_{197}$	trans-Neptunian object	890 km

[a] The objects listed here all have a diameter less than half that of Earth (that is, under 6,378 kilometers). All but the last three have a diameter greater than 1,000 kilometers, which, according to one proposed definition of a planet, would qualify them for planetary status.

[b] Note that in comparison to the larger of the planetary satellites Pluto is a small object (no fewer than seven planetary satellites are bigger than Pluto), and its diameter is less than half that of the smallest planet, Mercury. At the same time, Pluto is considerably larger than other known trans-Neptunian objects. Prior to the discovery of the TNOs, in the 1990s, the largest asteroid, Ceres, was the second largest of the known objects in the solar system. With an estimated diameter of 914 kilometers, Ceres is little more than a third the diameter of Pluto.

teenth century, before being reclassified as an asteroid, or a "minor planet"—a term that was coined on the occasion. That said, there are those who campaign for the restoration of planetary status to Ceres, which is by far the largest and most massive object in the asteroid belt. In the same way, there are people who would prefer to see the Earth-Moon system classified as a double planet rather than as a planet and its moon.

The main problem is that no formal definition of a planet exists. Various criteria have been proposed, but each stumbles over the problem of Pluto—either because Pluto would be included or because it would not, with each alternative being unacceptable to a significant fraction of astronomers. Nor is having a large satellite a helpful criterion because many TNOs are now known to be double. The only criterion that everyone can agree on is that a planet is a large body in solar orbit—but then, of course, nobody can agree on how to define the word *large*. One possibility is to consider any object greater than 1,000 kilometers in diameter in orbit around the Sun (that is, not a satellite of another body) as "large," in which case Pluto is a planet. Another is to set a fraction of the Earth's mass as the deciding factor—perhaps 1 percent, although numbers as high as 4 percent have been proposed. This latter criterion would exclude Pluto as a planet, but Mercury would be well inside the cutoff point.

Of these two definitions, the former is probably the more popular and may yet be adopted by the International Astronomical Union. But it has two major inconveniences. For one thing, if we describe anything larger than 1,000 kilometers as a planet, Pluto's status would be secured, but at least four other TNOs would have to be classified as planets as well: Quaoar, 2002 LM$_{60}$, (28978) Ixion, and the recently discovered 2004 DW (now officially named "Sedna"), which may even be only slightly smaller than Pluto. In addition, (20000) Varuna and 2002 AW197 may also exceed 1,000 kilometers, for a total of six. Given the rate at which TNOs are being discovered, within twenty years there may be dozens of them with diameters greater than 1,000 kilometers, and these would also have to be classified as planets, on top of which many astronomers regard it as inevitable that TNOs will be found that are as large as, or even larger than, Pluto. Second, Ceres poses a problem. Its diameter is currently estimated to be 914 kilometers, but some reference volumes give figures as large as 1,003 kilometers and 1,023 kilometers. If these larger numbers were correct, they would make Ceres a planet too. Does this then mean that as new measurements of its diameter are made, Ceres should be promoted to or downgraded from planetary status? Such an idea seems ridiculous.

So what is Pluto? It is not a comet because it does not develop a visible coma and tail. It is not an ordinary asteroid because it is in the wrong part of the solar system—although, at least for now, it is certainly king of the TNOs. Perhaps in the end all we can safely say is that Pluto is Pluto. I would, moreover, wager that more people are aware that Clyde Tombaugh discovered Pluto than remember that it was Johann Galle and Heinrich D'Arrest who discovered Neptune, and to this extent Clyde Tombaugh's legacy is safe.

Is Pluto an impostor? That's for you to decide.

SUGGESTIONS FOR FURTHER READING

Popular Books

Clyde Tombaugh and Patrick Moore, *Out of the Darkness* (Harrisburg, Pa.: Stackpole Books, 1980).

> *Clyde Tombaugh's own story of the discovery of Pluto, with supporting chapters by Sir Patrick Moore. A wonderful story and a wonderful book. Tombaugh ends his account when the search for Pluto was finished and thus avoids having to talk about his own exit from Lowell Observatory. After serving in the armed forces teaching navigation during the last two years of the Second World War, Tombaugh was made redundant for no readily understood reason on his return to Lowell.*

Mark Littmann, *Planets Beyond*, Wiley Science Editions (New York: John Wiley and Sons, 1990).

> *A thoroughly researched study that deals with the discovery of Pluto and the problems it presents. The book also offers an in-depth look at Neptune, Planet X, and many other topics pertaining to the outer solar system.*

Herbert Hall Turner, *Astronomical Discovery* (Berkeley: University of California Press, 1963 [1904]).

> *A fascinating book, dedicated to Edward Emerson Barnard by a colleague who worked with him during a visit to Yerkes Observatory. Turner tells the story of the discovery of Uranus and Neptune in detail and provides a fascinating analysis of the discoveries. A detailed, lucid, and beautifully written account.*

Looking around Our Solar System

More Advanced Reading

Dale P. Cruikshank and David Morrison, "Icy Bodies of the Outer Solar System," in J. Kelly Beatty and Andrew Chaikin, eds., *The New Solar System*, 3rd ed. (Cambridge, Mass.: Sky Publishing, 1990).

An excellent article that deals with Pluto in the context of the outer solar system.

On the Internet

The Cambridge Conference Network
http://abob.libs.uga.edu/bobk/cccmenu.html

The CCNet's archive of articles—drawn from a newsletter founded in 1997 to which many prominent science writers (including Arthur C. Clarke) have contributed—is a fabulous resource on many subjects. A series of contributions in 1999 and 2000 explore some of the issues raised by Pluto.

Pluto, the Ninth Planet
www.lowell.edu/users/buie/pluto/pluto.html

Marc Buie is one of the world's leading experts on Pluto and an active researcher. Although this is not an extensive Web page, it offers interesting information and useful links.

Clyde Tombaugh
www.klx.com/clyde/

A simple page on Clyde Tombaugh, the discoverer of Pluto.

CHAPTER 8

How Astronomers Learn without Going Anywhere

Many of our theories depend on learning things about planets (and stars and galaxies) without actually visiting them. Long before 1970, when the Venera 7 probe landed on the surface of Venus and measured its temperature, radio measurements taken from Earth in the late 1950s and then from the Mariner II spacecraft in 1962 had shown that it was bakingly hot—somewhere around 430°C. These measurements were made tens of thousands of kilometers from the planet itself. Then, in 1964, again without actually landing on the planet, Mariner IV told us that the atmosphere of Mars was extremely thin. We have a good idea of the surface composition of Pluto even though no space probe has been within a hundred million kilometers of the planet. And we know the composition of the Sun and the stars in great detail even though we have obviously never set foot on them. We can even make a pretty fair guess at the internal structure and thus the history of Mercury without a space probe ever having landed on the planet.

Just what tricks of the trade allow astronomers to deduce things about planets and other bodies that we have never visited? How is it that we are so often able to get things dead right (and sometimes so embarrassingly wrong)?

An Astronomer's Weapons

Until the late nineteenth century astronomers had only one weapon for studying the sky: their own eyes. Up to about 1880 all fundamental astronomy was carried out either by calculation or by looking through a telescope. Everything depended on the astronomer's skill, dedication, and eyesight. Telescopes were almost always small, and they were usually located in big cities. The two largest telescopes in existence in 1880 were William Herschel's 1.2-meter telescope at Slough, in England, which was finished in 1789, and Lord Rosse's 1.8-meter telescope at Birr Castle, in Ireland, finished in 1845. Neither Herschel nor Rosse was a professional astronomer, and their telescopes were ones they had built themselves, for their own studies.[1] At that time, it was often the amateurs—or at least the wealthy, and usually aristocratic, amateurs—who were better equipped and made the fundamental discoveries.

Two Great Advances

Two advances heralded the end of the golden age of the amateur astronomer and changed forever the way that astronomy would be carried out, leading to the present age of giant telescopes and sophisticated electronics. The first was the photograph. When photography was invented in the 1840s, some astronomers thought that it would revolutionize astronomy almost in an instant. It was proposed, for example, that with the benefit of photographic images a detailed map of the Moon could be prepared in a single night. Of course, such suggestions were absurd. Even under perfect conditions, the changing aspect of the Moon through its phases would make it impossible to produce a photographic lunar atlas in one night. (In fact, the first detailed photographic lunar atlas was not prepared until more than a century later, in the 1970s.) The problem was that early photographic plates had so little sensitivity and were so difficult to use that they had little practical application. By the 1880s, however, photography was finally ready to make its entry as a serious aid for the astronomer. The famous Harvard plate collection was begun in the 1880s, and many years later the first-ever observation of a quasar was found to have been recorded in a photograph exposed in 1888. The first good photograph of a comet was obtained in 1882, and on December 20, 1891, the first asteroid to be discovered photographically, (323) Brucia, was located by Max Wolf. By now, photographic plates were sensitive enough that for certain purposes they could replace the human observer. Unlike an astronomer, a camera does not get tired and lose its concentration and visual acumen, and a good camera is capable of registering stars far fainter than the human eye can see.

The second great advance had actually been made in 1665, by Sir Isaac Newton. At the time, Cambridge University had closed down on account of the epidemic of plague, and so Newton was at home in Woolsthorpe, in Lincolnshire, when he made his famous experiment. Using a prism to refract the light of the Sun, Newton was able to produce the rainbow of colors, thereby inventing the science of spectroscopy. He did not attempt to do anything further with his discovery, however, and it languished for nearly 150 years. In 1802 another Englishman, William Hyde Woolaston, discovered that dark lines ran across the solar spectrum, although he thought they were simply the boundaries between colors. It was not until 1814 that Joseph von Fraunhofer began the first systematic study of the lines that would eventually bear his name and produced a map showing the 324 such lines that he could detect in the solar spectrum. With the discovery—in 1859, by Gustav Kirchhoff—that each Fraunhofer line was caused by a particular element in the Sun and that the spectrum produced by each element had its own characteristic set of lines, as distinctive as a fingerprint, spectroscopy was ready to become a formidable research tool. Astronomers now had a means by which to analyze the light of distant planets and stars and thus to work out their chemical composition in detail.

It was only a matter of time before someone would take the next logical step and combine photography and spectroscopy, thereby opening the path along which astronomy would advance for almost a century to come. In 1889, Henry Rowland, the first professor of physics at Johns Hopkins University, produced a detailed photographic atlas of the solar spectrum. This atlas marked the imminent death, for professional astronomers, of visual astronomy and the beginning of a fundamental change in telescopes and observatories.

Bigger and Better

Until the end of the nineteenth century professional telescopes were generally small. We have already seen how Astronomer Royal George Airy was concerned that the largest instrument at the Royal Observatory at Greenwich, a 17-centimeter refracting telescope, would be inadequate to the task of locating Neptune and so delegated the search to James Challis, at Cambridge. This was typical of the time. Although the Cambridge observatory boasted a 30-centimeter reflector, this paled in comparison to Sir William Herschel's 1.2-meter and Lord Rosse's 1.9-meter reflectors. It was also typical that the observatory at Greenwich, although sited on a small hill in what had once been a small village at some distance from London, at the end of the nineteenth century found itself in what was swiftly becoming a suburb of London, increasingly hidden under smoke and soot from the rapidly spreading city.

It was a much maligned astronomer, Percival Lowell, who was to lead to the next great change. Lowell wanted to be able to observe the planets, Mars in particular, in unprecedented detail. Lowell realized that this could not be done from city observatories, which had to contend with dirty air and artificial light, and that an isolated mountain, preferably a fairly tall one, would offer much better viewing conditions. In 1894, he founded the first great mountaintop observatory in Flagstaff, Arizona, at an altitude of 2,210 meters. Unfortunately, Lowell Observatory was never wealthy enough to keep pace with the second step: that of placing ever-larger telescopes on these mountain sites.

The end of the nineteenth century saw a wave of construction of larger telescopes, but, even so, none approached the size of Lord Rosse's telescope at Birr Castle, near the village of Parsonstown. The first great modern reflecting telescope, the 1.55-meter telescope at Mount Wilson, completed in 1908, was still well short of the size of Lord Rosse's aptly nicknamed "Leviathan of Parsonstown." However, the Mount Wilson telescope was still by far the largest professional telescope and, even though the mountain on which it was built would, at 1,742 meters, be scorned today as too low for a major observatory, the rapidly developing techniques of spectroscopy and photography combined to produce astonishing results. At the same time, astronomers realized that they would need more light if they wished to observe and analyze ever more distant objects in greater detail. In 1917, the 1.55-meter reflector at Mount Wilson was joined by a new 2.5-meter telescope, the Hooker reflector, which would remain the world's largest telescope for more than thirty years. It was eventually superseded by the brilliantly successful 5.08-meter telescope at Mount Palomar, which became operational in 1948, and the less successful and underused 6-meter reflector at Zelenchutskiya in the former Soviet Union, which was completed in 1976 but suffered from a defective mirror and an assortment of other problems.

Now, even famous observatories like Mount Palomar are no longer regarded as good enough for the new generation of modern telescopes. For many years it was thought there would never be a successful working telescope larger than the famous 5.08-meter at Mount Palomar, but the past few years have brought an astonishing change. When it opened in 1988, the 4.2-meter William Herschel reflector at the Roque de los Muchachos Observatory on La Palma, in the Canary Islands, was the third largest telescope in the world. Now, early in 2004, it is not even among the ten largest and its position in the list is dropping fast. A new generation of telescopes that are 8, 10, and even 12 meters in diameter have been entering service. Mauna Kea, in Hawai'i, at an altitude of 4,205 meters, is the site of the two 10-meter Keck telescopes, at the William M. Keck Observatory, plus the 8.2-meter Gemini North and Subaru telescopes. Similarly, in northern Chile, not too far outside the city of La Serena, Cerro Pachón boasts the Gemini South Telescope, also 8.2 meters in diameter, as well as the four telescopes that together form the VLT, for "Very Large Telescope." In the near future, the island of La Palma, at an altitude of 2,340 meters, will become the home of the 11.4-meter Gran Telescopio CANARIAS, or GTC. Nor are these the end to it. Negotiations are currently underway for the construction and installation of telescopes that will range from 25 to 100 meters in diameter. These include the 30-meter CELT ("California Extremely Large Telescope"), to be built by the University of California in partnership with the California Institute of Technology, while the Swedish-Spanish Euro50 project will produce a 50-meter telescope for La Palma. Almost incredibly, plans exist to construct a 100-meter telescope for the European Southern Observatory, aptly to be named OWL: Overwhelmingly Large Telescope. The final site is as yet undecided, but it will probably be somewhere in the Atacama Desert, in northern Chile.

Technologically such huge instruments are not yet possible, although they will be within the next five to ten years. But telescopes such as OWL will probably mark the absolute limit of construction on Earth, given the complexities of construction and the need to protect the telescope from the weather. These instruments are so tall that even the force of the wind will vary at different points in their height, because the atmospheric pressure will change significantly between the base and the top of the telescope. A telescope larger than OWL will therefore surely be an instrument to be installed on the Moon.[2]

Invisible Eyes

The modern astronomer no longer relies exclusively on optical telescopes. Rather, in making their observations, astronomers increasingly employ what is often termed "invisible astronomy." Invisible astronomy, as its name suggests, uses something other than visible light to study the sky. There are many types of invisible astronomy. One does not even use light at all but instead uses neutrinos, the ghostly particles emitted from the core of the Sun and from many other objects such as supernovas. For the most part, though, invisible astronomy works with forms of electromagnetic radiation similar to visible light but of a longer or shorter wavelength.

So what are these types of invisible astronomy?

Radio Astronomy

Radio astronomy is the oldest method of invisible astronomy. It was developed in the 1920s but had its golden age after the Second World War. Radio waves are essentially like visible light, except that the wavelengths that make up visible light are extremely short—fractions of a micron (a millionth of a meter). In contrast, radio waves are measured in centimeters and even meters, which is why very large radio telescopes are required to detect them. Radio telescopes are equipped with receivers not unlike those in a home radio, albeit millions of times more sensitive. Although, of the bodies in the solar system, only the Sun and Jupiter are strong emitters of radio energy, the manner in which radio waves move through a planet's atmosphere furnishes important information about that atmosphere, particularly its density and structure. For example, when a space probe passes behind a planet, radio telescopes can be used to measure the effect that the planet's atmosphere has on the signals from the probe. This was how, in 1964, Mariner IV made the first accurate determination of the atmospheric pressure on the surface of Mars. The closely related discipline of radar astronomy—RADAR originally stood for "RADio location And Ranging"—has likewise been fundamental in the exploration of the solar system. Many objects, including the Sun, Mercury, Venus, the Moon, Mars, several Earth-approaching asteroids, and even a few comets have been examined by radar, which detects objects by bouncing radio waves off them, the same way that we use the beam of light from a flashlight to locate someone in the dark. Radar has proved critical in mapping Venus, in finding adequate landing sites for the Viking probes to Mars, and for exploring the size, shape, and even texture of objects that pass close to the Earth. Radar has also been used with great success to track meteors in the Earth's atmosphere and even to identify major meteor showers that take place in broad daylight.

Infrared Light

Infrared radiation is often termed *heat radiation* or, more correctly, *thermal radiation*. All bodies emit thermal radiation in accordance with their temperature. The hotter the object, the shorter the wavelength and the bluer the light. We can see this effect in the colors of the stars: the visible surface, or photosphere, of a blue star is about 25,000°C, whereas the surface of a red star is only about 3,000°C. Cooler objects, such as planets, also emit thermal radiation. The Earth emits infrared radiation at a wavelength of approximately 10 microns, which is some twenty times too long to be visible with our eyes. Infrared radiation contains a huge amount of information about the objects that emit it. Apart from giving us the temperature of an object, many atoms and molecules have lines in the infrared spectrum similar to the Fraunhofer lines in the solar spectrum that permit them to be identified and the composition of celestial objects thus analyzed.

We are able to observe infrared light by using one of three special semiconductor materials that convert electromagnetic radiation—light—into weak electrical signals. The material used depends on the wavelength: mercury cadmium telluride (HgCdTe) for wavelengths ranging from 1 to 2.5 microns, indium antimonide (InSb) for the 1- to 5-micron range, and indium arsenide (InAs) for wavelengths from 8 to 25 microns.

Ultraviolet Light

Ultraviolet radiation has a shorter wavelength and is more energetic than visible light. Ultraviolet light is what causes burns when our skin is exposed to sunlight for too long, but fortunately it is largely blocked by the Earth's ozone layer. Only the Sun is a significant emitter of ultraviolet radiation in our solar system, although by observing planets such as Venus and Mars in ultraviolet light (reflected from the Sun) we are able to gather important information about their atmospheres and clouds. Their atmospheres absorb ultraviolet light, but the light penetrates to different depths according to the atmosphere's composition. The polar regions of Jupiter also emit some ultraviolet radiation from auroral activity.

Most types of glass absorb ultraviolet light and mirrors swiftly lose their reflectivity. For example, a typical mirror covered with aluminum is quite a good reflector of ultraviolet light, but when exposed to air it rapidly becomes covered with a thin layer of aluminum oxide, a substance that does not reflect short wavelengths. The aluminum oxide deposits thus rapidly reduce the mirror's effectiveness, making it necessary to use special mirror coatings such as magnesium or lithium fluoride to protect the aluminum. Ultraviolet radiation is detected by the use of special phosphors that fluoresce, turning ultraviolet light into visible light. Although this may sound complicated, anyone who has been in a discothèque will probably have noticed that when people enter the beam of an ultraviolet light, certain items of their clothing start to shine—that is, they fluoresce under ultraviolet light.

High Energies

Beyond the ultraviolet are a series of even shorter wavelengths—the extreme ultraviolet range, followed by X rays and then gamma rays, all of which we can group together as "high energies." These are of considerable importance in the study of the solar system, for they are produced naturally by radioactive elements, and by analyzing this radiation we can determine the surface composition of a planet. The bombardment of a planet's surface by cosmic rays also generates high-energy radiation. One important new technique, X-ray fluorescence, enables us to carry out a detailed assessment of a planet's surface composition by examining the "signature" of the X rays produced by this bombardment. Similarly, radioactive elements in rocks produce gamma rays that can be detected by space probes, which permits us to work out the composition of the rocks on the basis of the radioactivity they emit. These high-energy photons are detected using a variety of cutting-edge techniques such as X-ray telescopes equipped with what are called "grazing incidence" mirrors. X rays are not normally reflected by mirrors: they are reflected only when they strike the mirror at a glancing angle. An X-ray telescope contains a series of nested, almost vertical mirrors that allow the X rays to land at the appropriate angle and reflect to form an image. In fact, such telescopes are able to produce high-resolution images of about the same quality as a crisp photograph.

High-energy astronomy is changing rapidly year by year, as increasingly sensitive detectors become available. When one moves into the range of gamma rays—light that is even more energetic than X rays—it is no longer possible to form photographic images using mirrors. Instead, a number of other techniques can be used to detect gamma rays and to create some kind of image from them. Primitive devices such as spark chambers, which consisted of rows of wires under high tension that would literally spark when hit by gamma rays, have evolved into modern semiconductor devices—relatives of the CCDs now so often found in electronic equipment. These devices are used in combination with a metal mask punched through with a pattern, which rotates in front of the detector, thereby either allowing the incoming gamma rays to pass or blocking them out. In this way, the stream of incoming rays is converted into a coded signal (much like Morse code). A sophisticated computer analysis of the signals produced by the semiconductor is then used to calculate precisely where each individual photon originated, which allows high-resolution images to be constructed from the signals. (The more complicated the pattern on the mask, the more detailed the information encoded in the signal, and therefore the better the computer-generated image.) These techniques allow the elements in distant stars and supernovas to be identified by means of the characteristic pattern of gamma rays they produce.

How Hot Is That Volcano?

As I write, one of my students is wrestling with a fascinating research problem that she has chosen for her Diploma of Advanced Studies (the Spanish equivalent of a master's degree). She wants to measure the heat of a volcano located on one of Jupiter's moons, more than 600 million kilometers away, and, on the basis of that measurement, understand how the volcano works. Io, the closest of Jupiter's four large Galilean satellites, is the most volcanically active object in the solar system. The Voyager encounters identified thirteen active volcanoes on Io, and no fewer than eight simultaneous eruptions could be seen in the images returned by Voyager I. Now, more than two hundred calderas—that is, volcanic craters—have been identified, sixty-one of which are active.[3] By the decision of the International Astronomical Union, Io's volcanoes are known by the term *patera*: Loki Patera, Surt Patera, and so on. Io's volcanoes, though, are not like volcanoes on Earth.

When Io was photographed by Voyager I, the first thought that came to the minds of many scientists was that it resembles a giant pizza. The satellite is a riot of color: the red of the tomato sauce, the white of the mozzarella cheese, the black spots of the olives. But these colors provide a clue to what is happening on the surface of Io. All these colors are typical of sulfur. One of the classic experiments in high school chemistry is to heat a quantity of sulfur in a test tube over a Bunsen burner. Sulfur crystals are normally a beautiful primrose yellow, but if they get extremely cold, they lose their color and become almost white. Heat them, however, and as they melt they turn golden. As the liquid sulfur heats further, the color intensifies and passes through orange to a deep red color. The liquid sulfur becomes almost solid again as long chains of sulfur atoms form, before the heat causes these chains to break up. The sulfur becomes nearly black and turns liquid again, and finally boils. What the colors of Io are telling us is that there is a lot of sulfur on the surface of Io and that this sulfur exists at widely differing temperatures. There are even lakes of molten sulfur, at temperatures of several hundred degrees centigrade, that fill many of Io's volcanic calderas, some of which appear to have rafts of solid sulfur floating on them.

Io is a crazy world that is rapidly turning itself inside out. It is estimated that in just a million years the equivalent of the entire interior of Io gets blown out through its volcanoes. Many things about Io, however, are not well understood. Scientists initially thought that all Io's volcanoes expel molten sulfur, but they soon realized that this was not possible. The plumes from the volcanoes that can be observed are produced by sulfur dioxide gas and are much cooler than molten sulfur. The Galileo mission identified hot spots—that is, volcanoes—on Io that have a temperature around 1,800°C, far too hot to be produced by sulfur, which boils on Earth at 444.6°C and, in the vacuum of Io's surface, boils at a temperature about 200°C cooler. Some eruptions, at least, must involve ordinary lava, although in many cases this lava is far hotter than any lava on Earth. We thus have at least three types of volcano on Io: volcanoes that blow out sulfur dioxide gas, volcanoes that emit liquid sulfur, and volcanoes that spew out ordinary lava made from melted rock.

How many of each kind of volcano are there? Exactly how hot are they, and how much does the temperature change during an eruption? How often does each volcano erupt, and how violent are the eruptions? These are some of the questions that astronomers would like to answer. Up to now we have had only the two Voyager flybys, each of which was able to have a close look at Io for just a single day. Although the Galileo spacecraft also made a number of passes close to Io, it suffered from a crippled antenna, with the result that only a tiny fraction of the images that had originally been planned could be sent back to Earth.[4] Soon we will be able to do better, however, at least in certain respects. Io's volcanoes can be observed directly from Earth, but up to now not clearly enough to supply the answers that astronomers are after. Why is this, and how can we do better?

Despite its volcanoes, Io is a very cold world. Much of the surface has a temperature around 150°C. The volcanoes, though, are hot and emit a lot of infrared radiation. Hold your hands to an oven, and you will feel the heat on your hands. What you are doing is detecting the infrared radiation the oven produces. Using a large telescope and a sophisticated infrared detector we can "feel the heat" from Io's volcanoes. According to its temperature, a volcano will emit infrared radiation at a certain wavelength, roughly between 3 and 10 microns. If we look at Io in the infrared, we'll see a hot spot (the volcano), superimposed on a cold, dark background. The problem is that Io is a small world, slightly smaller than our Moon, and even when it is closest to us, it is still 620 million kilometers from the Earth. Even in images taken with the Hubble Space Telescope it is small, which of course makes it difficult to study. The best mid-infrared views obtained to date had a resolution of 0.6 arcseconds, but Io itself is only 1.2 arcseconds across.

New giant telescopes such as the 10-meter Keck telescopes in Hawai'i and Spain's 11.4-meter Gran Telescopio CANARIAS (GTC) are in the process of installing sophisticated new infrared cameras that will enable us to study the sky in unprecedented detail. Because the atmosphere is steadier when we observe in the infrared spectrum, the only real limit to the sharpness of images is the size of the telescope itself. With enormous telescopes like the future 50-meter Euro50 and the 100-meter OWL, we will be able to obtain extremely detailed views of Io in

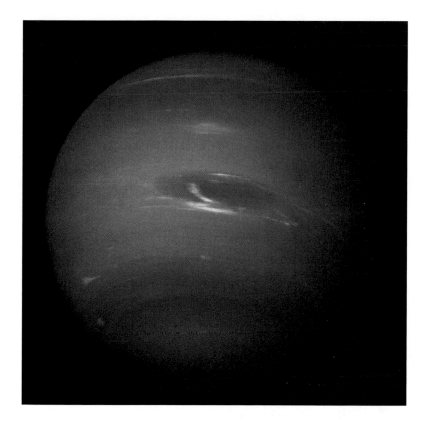

Neptune's Great Dark Spot (*center*) is as large along its longer dimension as the Earth.

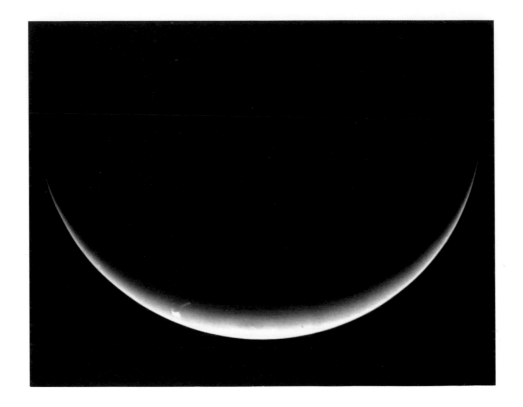

Neptune's south pole

This wide-angle image, taken through the camera's clear filter,
is the first to show Neptune's rings in detail.

Neptune and Triton (*foreground*)

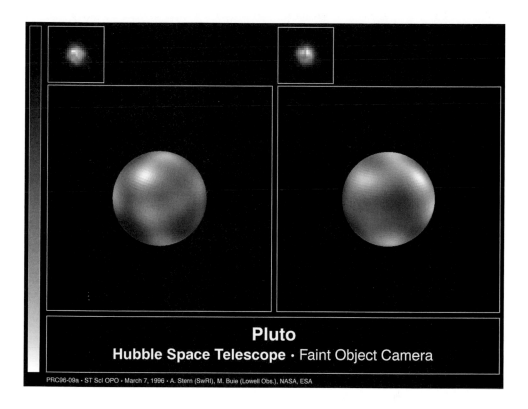

Pluto
Hubble Space Telescope · Faint Object Camera

PRC96-09a · ST ScI OPO · March 7, 1996 · A. Stern (SwRI), M. Buie (Lowell Obs.), NASA, ESA

The never-before-seen surface of the distant planet Pluto is resolved in
these NASA Hubble Space Telescope pictures.

Jupiter's moon Io

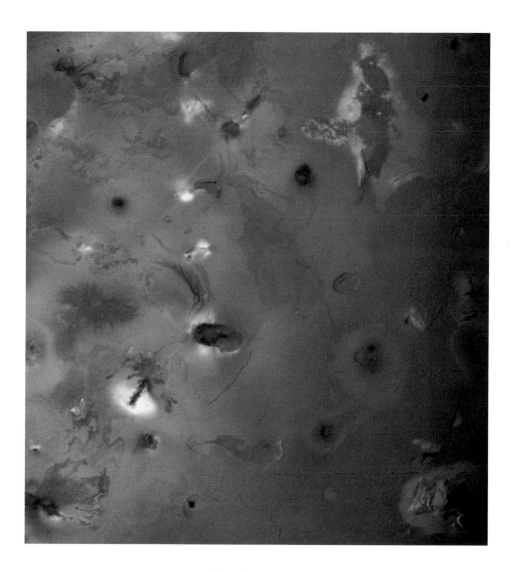

South polar region of Io

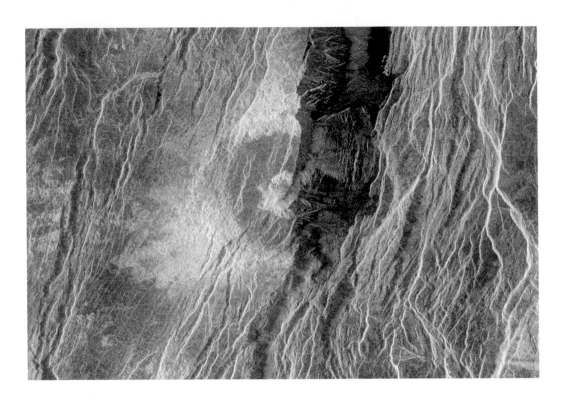

Somerville Crater in Beta Regio on Venus

Venus's Sapas Mons, a massive shield volcano

This image taken by the Halley Multicolour Camera on board the European Space Association's Giotto spacecraft shows the nucleus of Halley's comet as it approached the Sun in 1986.

An artist's conception of Deep Space 1's encounter with Comet Borrelly

The Arizona Meteor Crater

Copernicus, a crater located within the Mare Imbrium Basin on the northern
nearside of the Moon

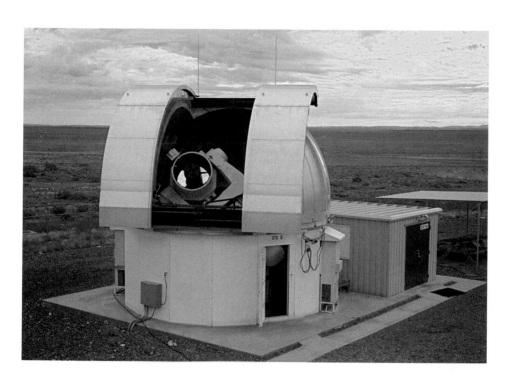

The Lincoln Near Earth Asteroid Research (LINEAR) Telescope. More than half of all the asteroids known to cross the orbit of Mars have been discovered with this telescope.

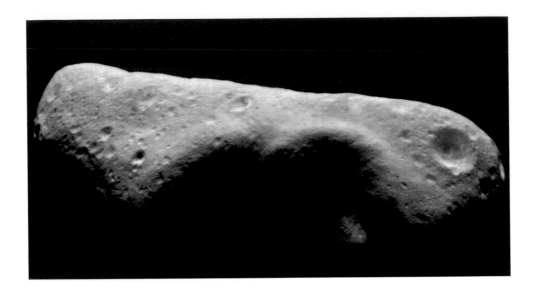

Image of Eros, an asteroid that was discovered by Rafael Ferrando. Ferrando realized early on that the asteroid was on a path heading straight for Earth.

The Experimental Test Site at White Sands, New Mexico, home of
the LINEAR Telescope

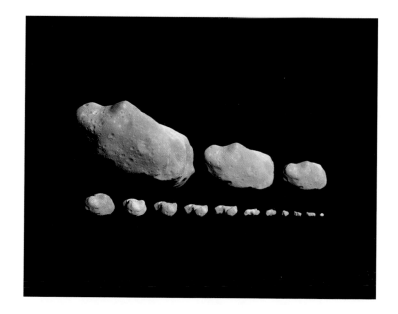

Asteroid Ida

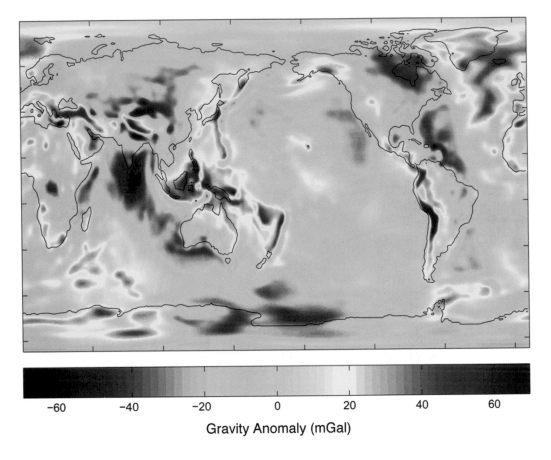

Gravity Anomaly (mGal)

Earth's gravity field as composed by the Gravity Recovery and Climate Experiment

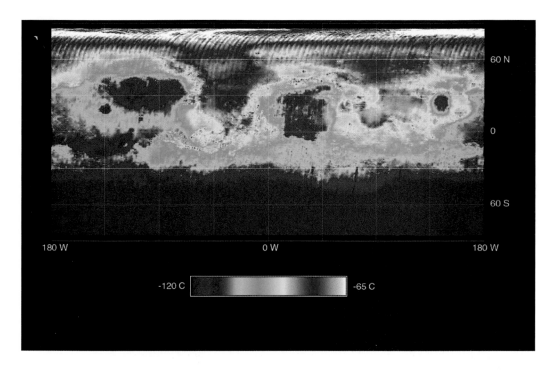

60 N

0

60 S

180 W   0 W   180 W

-120 C   -65 C

Nighttime Martian surface temperature

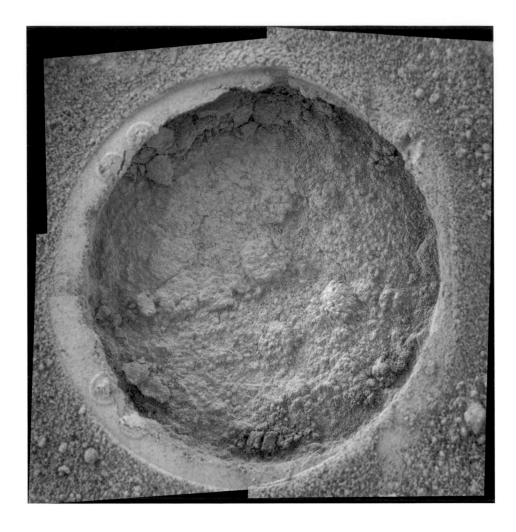

A target dubbed "Campbell" on a rock called "MacKenzie"
in Endurance Crater on Mars.

Malaspina Glacier in southeastern Alaska. Some experts predict that global warming may cause snow to disappear from many of Earth's mountains and glaciers within a few years.

View of the Grand Canyon acquired by the Advanced Spaceborne Thermal Emission and Reflection Radiometer (ASTER) instrument aboard the Terra spacecraft. ASTER will provide scientists with critical information for surface mapping and monitoring dynamic conditions and temporal change.

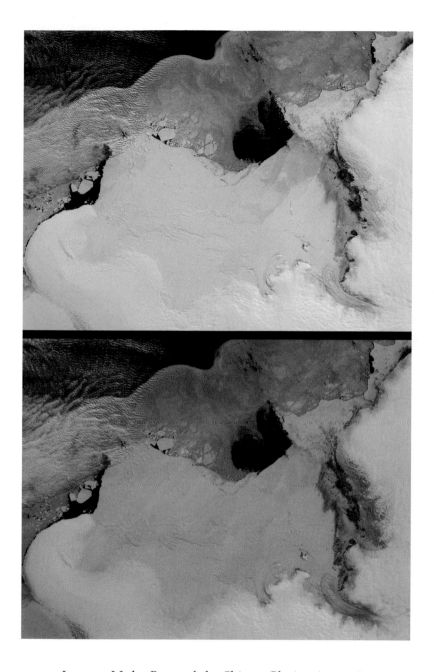

Lutzow-Holm Bay and the Shirase Glacier, Antarctica

A group of aligned barchan
sand dunes in the Martian
north polar region

Launch of Mars Pathfinder from Cape Canaveral Spaceflight Center
on December 4, 1996

Launch of the Lunar Prospector from Cape Canaveral Spaceflight Center
on January 6, 1998

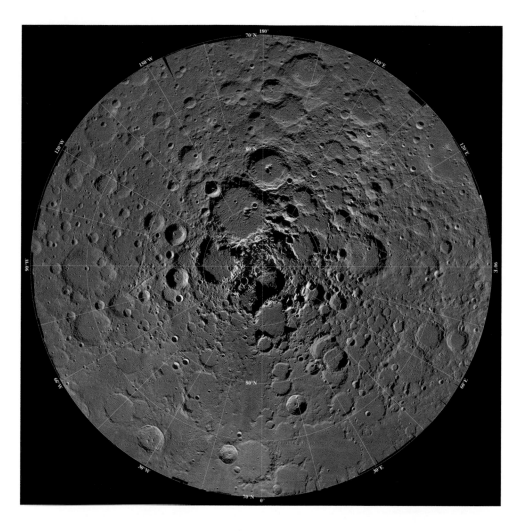

Photo mosaic of the north polar region of the Moon. The Moon's polar regions are of special interest because of the postulated occurrence of ice in permanently shadowed areas. The north pole of the Moon is absent the rugged terrain seen at the south pole.

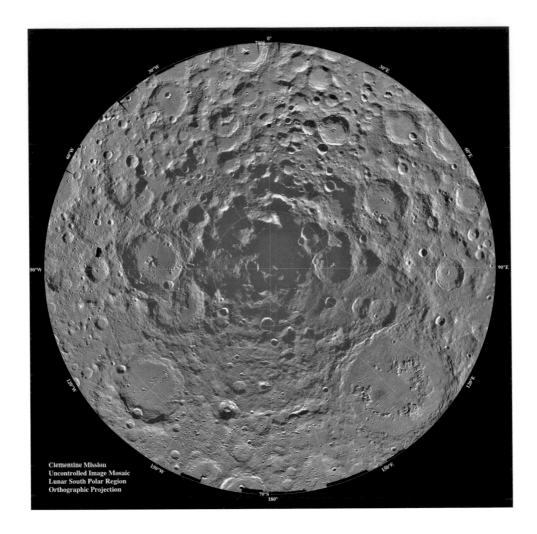

The south polar region of the Moon. The projection is orthographic, centered on the south pole. The south pole is of greater interest than the north because the area that remains in shadow is much larger than that at the north pole.

The heart of the Crab Nebula

Triffid Nebula

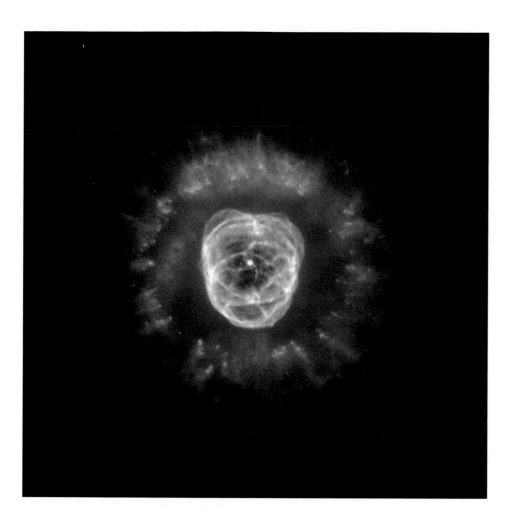

The "Eskimo" Nebula

Butterfly Nebula (*top*).
Cygnus Loop Supernova
blast wave (*bottom*).

An artist's conception of a possible newfound planet, one theorized to be at least as massive as Jupiter and to have a similar appearance to what the giant planets in our own solar system looked like billions of years ago

The Moon illuminated solely by light reflected from the Earth

the infrared range. But even with a telescope such as the GTC we can resolve structures on the surface of Io as little as 600 kilometers in diameter—not as good as the Hubble but good enough to enable us to pinpoint specific volcanoes. Thus, with the amazing resolution of the GTC, it will be possible for my student to resolve, identify, and investigate individual volcanoes on Io. This cannot be done using the Hubble images because they capture only the light that Io reflects from the Sun, not the heat that each volcano emits, with the result that both the volcanoes and the individual eruptions themselves are invisible.

How does all this help my student measure the temperature of a volcano on Io?

Each volcano emits a certain amount of infrared radiation that varies with its temperature. For example, a very hot volcano will emit most of its radiation at a wavelength around 3 microns and at longer wavelengths will emit less and less. A much cooler volcano will emit only a little infrared light at 3 microns, rather more at 5 microns, a lot at 8 microns, and then the amount of radiation emitted will drop off again as the wavelengths grow even longer. The amount of radiation emitted by the volcano at different wavelengths produces what is known as a "black body curve" (so called because the radiation emitted by a perfectly black object will generate exactly the curve that mathematical theory predicts). The curve rises sharply at the beginning and then comes to a peak at a wavelength that corresponds to the temperature of the volcano, after which it falls off more gradually. Thus, if we measure the amount of infrared radiation that a volcano emits at a number of different wavelengths, we can establish where the peak is located and thus measure the temperature of the volcano. The more accurate and thorough our measurements, the more precisely we can determine the temperature, and the aim of my student is to employ certain special techniques that will enable her to obtain measurements that are more precise than anyone else in the world has succeeded in making.

Once we know how hot the volcano is, we can tell what type it is. In fact, we can do better even than that. If we measure the total amount of radiation that the volcano emits—in other words, the brightness of the volcano—we can measure how much "lava" of whatever type it is emitting. By observing a volcano over several days, we can measure the progress of an eruption almost as well as if we were vulcanologists hanging instruments over the side of the crater. Of course, it is a lot safer to do this from the comfort of a telescope control room 500 million kilometers from Io's raging hellfires and frozen deserts.

Seeing through the Clouds of Venus

Long before anyone imagined that a probe could ever land on Venus, radio observations from the Earth, and then from the Mariner II spacecraft, were able to show that the planet's surface temperature was somewhere over 400°C. The method used in these observations was similar to the one my student will use to measure the temperature of volcanoes on Io. We also know that Venus has active volcanoes. But how do we know this when nobody has ever seen a Venusian volcano and probably never will? Or how, for example, do we go about measuring the rotation period of the planet?

During the 1960s the techniques of radio astronomy advanced to such a degree that feats unimaginable just a few years earlier suddenly became almost easy. In 1946, when the first radar echoes were received from the Moon, it seemed inconceivable that just fifteen years down the road we would be able to use radar not only to locate a planet, such as Venus, but also to map its surface. The construction of giant radio telescopes equipped with highly sophisticated detectors, such as those at Goldstone, in California, and at Arecibo, in Puerto Rico, made this possible. The procedure involved an adaptation of the same techniques used to detect the large amounts of microwave radiation coming from Venus, which had already led astronomers to speculate that the surface of Venus might be incredibly hot.

The story of Venus begins in 1956, the year in which the first measurements were made of the radiation that the planet emits. These measurements, taken by a radio telescope operating at a wavelength of 3.15 centimeters, indicated that Venus produces large quantities of microwave radiation, that is, radiation with wavelengths of only a few centimeters (the same radiation used in the microwave oven in your kitchen or by airport radar to track the flight of your airplane). If this radiation was thermal radiation—in other words, if it was due to the heat emitted by the planet—then the surface temperature of Venus must be around 350°C. Additional measurements made by other radio telescopes operating at different wavelengths substantially confirmed this result. Although alternative explanations for the presence of so much microwave radiation were suggested, most astronomers had little doubt that they were dealing with a very hot planet. Mariner II proved that they were right.

By 1961 techniques had improved to such an extent that for the first time

Venus could be detected by radar. This encouraged astronomers to try mapping the planet's surface. Radar mapping revealed certain persistent surface features, in particular two bright regions that were named Alpha and Beta. By mapping the movement of these two areas across the planet's disk, astronomers were able, in 1962, to make a first estimate of the rotation period of the surface of the planet itself, as opposed to the clouds that cover it—about 250 days, a number surprisingly close to the one accepted today. Over the years the period has been refined to 243.01 days, leading to the remarkable situation that the surface of the planet takes more than a Venusian year of 227 days to rotate once on its axis. Moreover, it does so backwards, so that the Sun rises in the west, not the east. In other words, if you were to view the planet from above its north pole, you would see that it rotates clockwise, whereas the Earth, the Moon, Mars, Jupiter, and so on all rotate counterclockwise. However, the clouds 60 kilometers above the surface rotate a very great deal faster, every 3.995 days, and in the opposite sense to the planet's surface.[5]

In December 1978 the Pioneer Venus probe entered orbit around the planet and carried out a detailed radar mapping of the surface. The mapping depicted a planet of great rolling plains and two large continents, which have since been named Ishtar Terra and Aphrodite Terra. The Pioneer Venus was followed by the Magellan probe, which began a radar mapping of Venus in September 1990 and detected astonishing structures—volcanoes, fractures, flat domes that resemble pancakes—on the surface of the planet. Both Pioneer and Magellan revealed another curious fact: 90 percent of Venus is within 1.5 kilometers of the nominal sea level (that is, the mean radius of the planet, or the average distance from the planet's surface to its center), making Venus for the most part a very flat planet. For Earth, the corresponding figure is about 35 percent and, for Mars, 30 percent. But if Venus consists mostly of flat plains, it also has towering peaks. For example, the range of mountains known as Maxwell Montes reaches 11.1 kilometers above sea level, putting Mount Everest, at a mere 8.8 kilometers, to shame.

At first, it was hard to understand how such a terrain came to be. There is no evidence that Venus has plate tectonics like the Earth. Rather, the whole of Venus seems to be a single plate. How can there be such high mountains in a localized region when the temperature of the surface of the planet is so high that rock will slowly ooze like thick treacle? Over millions of years any high mountains should gradually flatten out under their own weight as the rock slowly spreads across the surface of the planet. The obvious answer is that Venus has volcanoes that progressively build up new mountains to replace the old. Close examination of the Pioneer Venus and Magellan radar maps has confirmed this theory. In fact, Venus has more volcanoes than any other planet in the solar system. Over sixteen hundred major volcanoes or volcanic features such as domes of lava and lava flows have been identified, and there are many smaller volcanoes. No one has yet counted them all, but the total number may be over a hundred thousand—or even over a million.

The radar maps have thus shown us that Venus has a huge number of volcanoes, but what is the evidence for active volcanoes? How can we "see" active volcanoes when we cannot even see the surface of the planet through its dense covering of cloud?

There are two main pieces of evidence that, taken together, are convincing. The best evidence is offered by the changing levels of sulfur compounds in the atmosphere. On Earth, volcanoes emit sulfur dioxide, and in 1978 the Pioneer Venus probe measured high levels of sulfur dioxide and of sulfuric acid haze in the Venusian atmosphere. In fact, the level of sulfur dioxide was far higher than had been expected on the basis of observations of Venus from the Earth. By 1983, however, the amount of sulfur dioxide had fallen by about 90 percent. The second piece of evidence was provided by radio observations, both from the Earth and from Pioneer Venus. These detected weak, low-frequency radio emission from three areas: Beta Regio, a highland region that has two 4.5-kilometer peaks flanked with what appear to be lava flows, Phoebe Regio, which is not far from Beta Regio, and Atla Regio, another region that appears to have many volcanic structures. The radio emission from these areas is similar to the radio waves produced by lightning. We know that lightening discharges have often been seen in the plumes of volcanoes on Earth, so the emission of radio waves strongly suggests that these areas contain active volcanoes. In combination, these two pieces of evidence point to the conclusion that Venus suffered an extremely violent volcanic eruption in the late 1970s, sending huge quantities of sulfur dioxide into the atmosphere, and that these eruptions probably occurred in the Beta Regio area of the planet.

Our invisible eyes have thus been able to tell us a great deal about the surface of a planet that is permanently veiled in an impenetrable layer of cloud, to the point that we can even "see" the titanic eruptions of volcanoes.

The Invisible Nucleus

As a final example of what invisible astronomy can teach us, let us turn to the story of a race that has not yet been run to measure something that cannot yet be seen. It was only recently, in 1950, that Fred Whipple proposed to a not entirely convinced world that at the heart of each comet is an icy nucleus. But no one had seen this nucleus, and attempts to observe it were in vain. As we now know, the reason is that the nucleus is so small—in many cases only a kilometer, or at most a few kilometers, in diameter—that no telescope could ever view it directly, particularly when the nucleus is hidden in the middle of the comet's dusty coma. Only when a comet is a long way from the Sun and its activity is completely frozen out, so that the nucleus no longer emits the gas and dust that conceal it, can we hope to see what astronomers term "a bare nucleus." Of course, the nucleus of a distant comet is faint and thus difficult to observe. What is more, by way of an "et tu Brute," the nucleus is not the brilliant white of ice and snow. In fact, it is as black as coal—a dirty snowball indeed—and thus far fainter than anybody ever anticipated.

It is no surprise, then, that some astronomers were skeptical about the existence of this invisible iceberg. It seems astonishing now that one of the main objectives of the Giotto mission in 1986 was to prove that there is a nucleus at the heart of Halley's comet. Although by then the other evidence was becoming conclusive, the final proof could only be found in the form of images of the nucleus. Giotto was a brilliant success, as were the Soviet Vega probes and the two Japanese missions, Sakigake and Suisea. The two Vega probes photographed the nucleus from a distance of about 8,000 kilometers, and then Giotto captured it as it passed by at just 605 kilometers.[6]

Vega and Giotto proved that the nucleus really exists, even if it was not quite the bright, highly reflective snowball that Fred Whipple, and others, had envisaged. Then, in September 2001, the Deep Space 1 probe made a brilliantly successful pass by the nucleus of Comet Borrelly, revealing a nucleus that in many respects was surprisingly similar to Halley's nucleus. Finally, on January 2, 2004, the Stardust probe passed only 240 kilometers from the nucleus of Comet Wild 2, which turned out to have a peculiar, starkly cratered landscape totally unlike the surface of the nucleus of Halley or Borrelly. Data from the Giotto probe also indicated that the nucleus of Comet Halley is far larger than anyone had imagined. Prior to the Giotto flyby, the best estimate of the diameter of the nucleus was generally accepted to be about 5 kilometers, whereas Giotto revealed that the nucleus had a shape often compared to a lopsided potato or a peanut and actually measured about 18 × 8.5 × 8 kilometers. Moreover, rather than spinning smoothly on its axis as it orbits the Sun, it was tumbling violently, like a drunk acrobat who cannot keep somersaulting in a straight line. It is also, sadly, cocooned in the center of a dense cloud of dust, so much so that even when Halley's comet is at its closest to Earth, its nucleus remains invisible.

What, though, of other comets? When Comet Hale-Bopp appeared in 1997, astronomers knew that it was large. Unfortunately, we still don't know how large. Many estimates have been made, using different techniques, which have yielded estimates of the nucleus's diameter that range from about 25 kilometers to about 130 kilometers. Even at the smaller of the two extremes, Comet Hale-Bopp would be a large object, roughly double the size of Comet Halley. At the other extreme, it would be more than ten times the size of Comet Halley and one of the largest comets ever observed. It is slightly embarrassing that such an important parameter of such an extensively observed object is so hard for us to nail down.

In the case of Comets Halley and Borrelly, it was data from space probes that allowed astronomers to estimate their size. In addition, a few objects—but only a few—have passed close enough to the Earth to have been detected by radar. One of these was Comet Hyakutake, which has a diameter somewhere between 1 and 3 kilometers. However, the majority of the estimates we have of the sizes of nuclei have been obtained by indirect measurements. Typically, astronomers measure some property of the comet and then attempt to infer the size of the nucleus. One reasonably successful approach has been to measure the brightness of short-period comets—small comets that generally have periods under twelve years and therefore return much more frequently than Halley's comet—when they are at their greatest distance from the Sun and not very active. We are aware that as some comets get further from the Sun, their brightness reaches a minimum value and gets no fainter.[7] In these cases we assume that what we are seeing, at such a great distance from the Sun, is the brightness of the nucleus itself. If we can measure how reflective the nucleus is, we can estimate how large it is—although, given all the guesses and approximations involved, such measures can be accurate only to within a factor of two. In other words, if we say that a comet's nucleus has a diameter of 2 kilometers, we can be confident that the diameter is greater than

Looking around Our Solar System

I kilometer but less than 4 kilometers. This may not be very good, but it is better than nothing.

To determine more accurately how large the comet's nucleus is, we will need to know how reflective it is. And to do that we must turn to our invisible eyes once more. The method we will use is one that has proved brilliantly successful for estimating the diameter of asteroids. We know that this method will work, but so far no one has been able to use it with comets.

To understand why this is, let us look more closely at what is involved. An asteroid, and likewise the nucleus of a comet, reflects a certain quantity of sunlight depending on its size and its reflectivity. If we measure the amount of reflected sunlight we can make an educated guess at the object's size and then calculate its reflectivity, or we can guess the reflectivity and then calculate the size, but we cannot calculate both. Now suppose that a comet reflects 5 percent of the sunlight that falls on it. That means that the remaining 95 percent is absorbed, heating the surface of the nucleus. If we can measure how much heat the nucleus emits, we will know how much sunlight it has absorbed. And once we know how much sunlight it absorbs, we can calculate just how much it reflects. To measure how much heat the nucleus of a comet emits we must make our measurements in the mid-infrared range, at around 10 to 20 microns. This seems a simple exercise.

It is not.

In the case of an asteroid, we can make the necessary measurements when the asteroid is relatively close to the Earth and thus relatively bright. In the case of a comet, however, the measurements must be made when the comet is as far from the Sun as possible and completely inactive, so that we measure only the nucleus itself rather than the coma or a cloud of dust around the nucleus, which would give us a false value of the brightness. But when a comet is a long way from the Sun, it is not only faint but cold, and so the amount of heat it emits is tiny. Making measurements in the mid-infrared is difficult. To measure the minuscule amount of heat emitted by the nucleus we will need a large telescope equipped with an extremely sensitive detector. Up to now, no telescope has been sensitive enough to make such measurements. However, a number of very large telescopes, such as the two Keck telescopes, the two Geminis (North and South), the Japanese telescope Subaru, and the GTC, are currently being put into service that will have powerful mid-infrared detectors. These telescopes are as much as a hundred times more sensitive to mid-infrared radiation than, for example, the 3-meter

NASA Infrared Telescope Facility in Mauna Kea. Add to this the Spitzer Space Telescope (previously known as SIRTF, the Space Infrared Telescope Facility) satellite that was launched on August 25, 2003, and features some excellent infrared detectors, and we have a number of telescopes that will be competing to measure the size of a comet's nucleus.

Who will win this particular race? The smart money is on SIRTF, which has the advantage of operating in space, beyond the Earth's atmosphere, making its instruments much more sensitive. A very able team will be using the Gemini South, however, and will track Comet Hale-Bopp as it moves slowly away from the Sun near the south pole of the sky. If we are right about the size of Comet Hale-Bopp, it should be easy to measure the infrared radiation from its nucleus. The only problem is that, even though it is already beyond the orbit of Uranus, it is still quite active. Will its activity finally die down before it gets so far away from the Sun that it can no longer be detected? For a while, everybody assumed it would, but doubts have now arisen. If Comet Hale-Bopp is a very great distance from the Sun before it ceases to be active, we may need to wait to make our measurements until the giant Next Generation Space Telescope is launched, possibly in 2016. This telescope, which will be more than three times the size of the Hubble Space Telescope, will be by far the most sensitive mid-infrared telescope yet to exist. Here in the Canary Islands we are out of the race to measure Comet Hale-Bopp because it is too far south to be seen. Furthermore, the GTC will probably not be fully operational until about mid-2006, which means that other telescopes will have a head start over us. But there will still be comets for us to measure, and each one will add another small piece of knowledge to the jigsaw puzzle that, when complete, will resolve the many mysteries of comets.

Optical astronomy has by no means been superseded. But when an astronomer wants to find out something fundamentally new, it is more and more likely that the information will come, at least in part, from invisible astronomy. The three examples we have looked at go to show just how much astronomers depend on tools other than just visible light. As recently as 1975, the idea of measuring the heat from a comet's nucleus or from a volcano on Io would have been inconceivable, in part because we either did not even suspect or else were just not sure that the objects in question actually existed. Even our observations of volcanoes on Venus, which are a little older, would have seemed unthinkable just ten years earlier. What is astonishing is that we can make such discoveries and reach such conclusions simply by listening to radio static from Venus or looking at the colors of a sulfur pizza.

SUGGESTIONS FOR FURTHER READING

Popular Articles and Books

Patrick Moore, "Venus—By Radar," in *The Sky at Night 3* (London: British Broadcasting Corporation, 1970), pp. 117–19.

> *A short historical article that discusses some of the first radar maps of Venus. What makes Moore's essay especially fascinating is that it allows us to see how much the techniques we use and the results they produce, as well as our confidence in these results, have changed since December 1968, when the program was broadcast.*

Jean Audouze and Guy Israël, *The Cambridge Atlas of Astronomy*, 3rd ed. (Cambridge: Cambridge University Press, 1994).

> *A beautifully compiled and richly illustrated volume that does an impressive job of covering many of the topics raised in this chapter clearly and accurately, but without going into excessive depth.*

Gary Hunt and Patrick Moore, *The Planet Venus* (London: Faber & Faber, 1982).

> *A relatively slim volume, but full of details concerning the observation of Venus, from historical material through to telescopic observations and eventually to data from space probes. Gary Hunt worked on a number of the NASA missions and is thus an expert on space probes and space probe data.*

More Advanced Reading

Peter Cattermole, *Venus: The Geological Story* (Baltimore: The Johns Hopkins University Press, 1994).

> *An exquisitely researched book on the geological history of Venus. Includes details of investigations both from space and from Earth.*

Torrence V. Johnson, "The Galilean Satellites," in J. Kelly Beatty and Andrew Chaikin, eds., *The New Solar System*, 3rd ed. (Cambridge, Mass.: Sky Publishing, 1990).

> *An excellent study of the four satellites of Jupiter, full of beautiful photographs and cogent explanations.*

How Astronomers Learn without Going Anywhere   179

On the Internet

Astronomical Picture of the Day
http://antwrp.gsfc.nasa.gov/apod/ap960805.html

> *One of the most spectacular images taken by Voyager of an erupting volcano on Io, with a series of links to further information.*

Io
www.nineplanets.org/io.html

> *The Nine Planets Web site's page on Io. Many links, nice images, and much background information.*

Volcanoes on Venus
http://volcano.und.nodak.edu/vwdocs/planet_volcano/venus/intro.html

> *A page of information on volcanoes on Venus, including maps and images from the Volcano World Web site.*

Volcanoes of Venus
www.volcanolive.com/venus.html

> *Detailed information about the different types of volcanoes found on the surface of Venus.*

Deep Space 1
http://nmp.jpl.nasa.gov/ds1/
http://nmp.jpl.nasa.gov/ds1/images.html

> *The Deep Space 1 Web pages, including superb images returned from Comet Borrelly.*

PART 3

Triumph or Disaster?

THE HUMAN RACE HAS OCCUPIED SPACESHIP EARTH FOR some four million years. For most of that time we have taken our home for granted, and we have perhaps grown too confident in our own safety. Over the past several decades, though, our attention has increasingly been drawn to how fragile our planet is. Ever since the development of the atom bomb we have lived with the threat of nuclear holocaust, and now other potential sources of disaster have begun to occupy the scientific community. One uncomfortable possibility, with which astronomers in particular are concerned, is that events dramatized in films such as *Meteor*, *Deep Impact*, and *Armageddon* could one day come to pass. Of more immediate danger, though, is the fact that we have already used up a significant part of the Earth's limited resources and have all too often run roughshod over the planet's ecosystems. What will become of us when we run out of options? People who think that exploring the solar system is a waste of time and effort need only look as far as our nearest neighbors in space to learn what the Earth's future could look like. It is a lesson we ignore at our peril.

The colonization of the solar system or even of the Galaxy has, of course, long been a standard theme in science fiction. Over the years fans of *Star Trek* have eagerly followed the adventures of James T. Kirk and his successor, Jean-Luc Picard, and audiences flocked to see *Star Wars* and its sequels. But will we ever actually travel among the stars? One thing is certain: to reach the stars we must dare to take that first step to the planets. In this section we will delve into some of the realities of exploring the solar system, including the technology that will be needed to reduce the length of journeys to other planets. We will also discover why our first step to the planets is likely to begin not from the Earth but from the Moon.

All the same, going to the planets and beyond may still appear to be the quintessential fantasy. When we look at the sky at night, the stars seem impossibly remote—so inconceivably distant that it is hard to feel any real connection to them. And yet the expression "we are stardust" is far more literally true than most of us imagine. You, me, the whole human race, and even our planet are here only thanks to the death of massive stars long ago, stars that seeded the universe with the very stuff of creation. Thus, when we study the stars, we are studying our own distant origins—and, in the end, how we got here is surely the greatest enigma of all.

CHAPTER 9

Clear and Present Danger:
Will We Be the Next Dinosaurs?

Over the past few years I have been asked more and more often about the danger that the human race's tenure on Earth might end in the same way as that of the dinosaurs 65 million years ago. In fact, I recently gave a lecture on the subject that was effectively by royal command, after yet another asteroid scare hit the headlines, and some of my colleagues were contacted by members of the Spanish royal family, who were concerned about the reality of the danger. Several times over the past decade or so, for a few days or weeks, there has indeed appeared to be a small, but worrisome, risk that sometime in the foreseeable future the Earth might be hit by an asteroid large enough to kill many millions of people. Hollywood has played its part in fanning the public's fears about the threat posed by comets and asteroids. The 1979 film *Meteor*—based, loosely, on a study by the world-famous Jet Propulsion Laboratory—thrilled audiences with a tale about the impending destruction of the Earth. In 1998 *Deep Impact* attempted to convince the public that the Earth was in danger of colliding with a marauding comet, whereas in the action-packed but totally unrealistic *Armageddon*, also made in 1998, the villain of the piece was an asteroid the size of Texas that—despite being one of the biggest asteroids known to exist—had somehow been overlooked by negligent astronomers until it was just two weeks away from colliding with the Earth. As this last film helpfully reassured us, though, in the event of such a threat NASA and Bruce Willis would need only two weeks and a bit of brute force to sort out the problem. If you took *Armageddon* seriously, please read on.

The end of the world is, of course, a topic of perennial interest. For many years, the possibility that nuclear war might bring the human race to an apocalyptic close seemed very real. With the demise of the cold war, asteroids and, more recently, climate change have provided suitably horrifying alternatives. The latest Hollywood disaster film features a new ice age that suddenly seizes hold of our planet in less than a week. When scientists were asked whether the film was realistic, many were torn between guffaws of laughter and outright horror that the public might actually find such a sudden deep-freeze scenario believable.

As new asteroids are discovered and as we are able to predict the path of known ones more accurately, it has become clear that in any given year the odds against the Earth being hit by an asteroid large enough to do truly cataclysmic damage are in the millions to one. We also know that only very rarely in the past has an asteroid of significant size collided with the Earth. But when such an event could easily kill hundreds of millions of people, even tiny probabilities cannot simply be dismissed. For example (and as we will see below), in 2002 scientists announced that there is as much as a 1-in-300 chance that the asteroid (29075) 1950 DA, which is over a kilometer in diameter, could hit the Earth in March 2880. Viewed the other way round, this means that there is a 99.7 percent probability that the asteroid will merely pass close by the Earth, with nothing happening—and, as a cynic might put it, a miss is as good as a mile. All the same, that remaining 0.3 percent is not something that can be ignored. If this asteroid were to strike the Earth the consequences would be devastating.

So far, we have been lucky: never has the probability of an impact increased as we continue to study an object. Instead, further investigation of the orbit has always led to a rapid decline in the odds as the object's position is tied down with greater accuracy. What we have yet to witness, then, is the situation depicted in the science fiction novel *Lucifer's Hammer*, in which the likelihood that a particular comet will collide with the Earth steadily increases as astronomers calculate its orbit more accurately. As Senator Arthur Jellison comments to the head of the

Strategic Air Command: "The odds are now hundreds to one against. Used to be billions. Then thousands. Now it's only hundreds. It's a little scary."[1] But if thus far we have been spared having to cope with such a situation, sooner or later we will be faced with an impending collision—although the threat is unlikely to develop as suddenly as Hollywood scriptwriters imagine. Rather, we would probably be looking at a long, drawn-out crisis that would be all the more horrific because we would be certain of the outcome—the inevitability of an impact and its exact location on the Earth—at least a few days, and possibly as much as some years, beforehand. But unless we had been alerted to the problem several decades in advance, we would be unable to do anything to head off the disaster.

Our Dangerous Solar System

It is easy to see that our solar system is a dangerous place. Even with a small telescope we can observe that the surface of the Moon is pitted with craters of all sizes, the largest of which are as much as 300 kilometers in diameter. In the southern highlands the craters are packed so tightly together that some lie on top of others. The first space probes to take close-up photographs of the lunar surface showed that there are millions of smaller and smaller craters, down to the tiniest pits that their cameras could resolve. Scientists studying the samples of moon rock that the Apollo astronauts brought back found that even the grains of dust clinging to the rocks were pitted with craters so infinitesimal that they were visible only with microscopes. Here on Earth we are fortunate that our atmosphere protects us from much of the stuff flying around space of the sort that produced the smaller lunar craters. Only quite large rocks can penetrate our atmosphere, and there are, thankfully, not too many of these. What is more, although our Earth is bigger than the Moon, it is still a relatively small target and therefore not all that easy to hit.

Until 1964 we knew only that craters existed on the Moon. As space probes started to explore our solar system, however, astronomers were startled to discover that craters can be found almost everywhere. The first, heartbreaking image of a battered, crater-strewn Martian landscape returned by Mariner IV forever changed our view of that world. Ten years later, the last of the Mariner probes, Mariner X, revealed that the planet Mercury was similarly battered. Many scientists thought that Venus, with its dense atmosphere, would be safe from major impacts, but radar studies from Earth and from orbiting satellites have shown that it, too, has huge craters on its surface.[2] The Voyager probes subsequently demonstrated that the satellites of Jupiter, Saturn, Uranus, and Neptune are massively cratered.[3] One satellite of Jupiter, Callisto, is so densely pitted that, as computer studies have been able to demonstrate, the total number of craters never changes because each new crater that forms destroys an old one, much as in the southern highlands of the Moon. We also know that the asteroids photographed by space probes are covered in craters, and we can only assume that the surface of Pluto is similarly a landscape riddled with craters. Only the four planets formed of gas, Jupiter, Saturn, Uranus, and Neptune, have no craters—but then neither do they have a solid surface.

As space probe after space probe has revealed the bruised face of our solar system, we have slowly been forced to acknowledge that we move in an interplanetary coconut shy. The good news is that the worst of the bombardment happened some four thousand million years ago. The bad news is that the notion that the bombardment then came to a halt is somewhat optimistic.

Dateline Arizona, Time Unknown

A few years ago I went to the observatory at Mauna Kea in Hawai'i to make some observations. The return flight crossed the United States from Los Angeles to Atlanta, where I would change to a transatlantic flight. About an hour after we left Los Angeles I glanced out of the left-hand window of the aircraft at the bare Arizona desert below. Precisely at this moment the endlessly flat landscape was broken by a breathtaking sight. The Barringer Crater, or the Arizona Meteor Crater, is spectacular from the air, even from a height as great as 10 kilometers. The view of the crater was totally unexpected, and it made a huge impression on me. And yet this crater, although immense by earthly standards, is so small that, were it on the Moon, it would almost certainly not merit a name, nor would it even be visible without a powerful telescope.

We do not know exactly when the Arizona Meteor Crater was formed. The best estimates place its age at anything from twenty-five to fifty thousand years. Despite Indian legends that talk of an event that split the sky, it is almost certain that the crater was formed long before the first settlers of North America crossed the land bridge from Russia to Alaska. At the same time, we do know that even though the object that formed the crater was rather small—probably only some

Triumph or Disaster?

50 to 100 meters in diameter—the explosion produced by its impact must have been the equivalent of somewhere between 50 and 100 megatonnes. Quite probably nothing would have survived within a radius of dozens of kilometers around ground zero.

Dateline Tunguska, 1908

June 30, 1908, started like any other day in the Tungus region of Siberia. Then, as now, the area was only lightly settled. Reindeer herdsmen, living a nomadic life in tents, were the only witnesses to the events of that morning. Around 7:00 A.M., a blinding light suddenly appeared from the south. After crossing the sky rapidly, an object of some sort exploded with a force of some 30 megatonnes. Seismometers around the world recorded the blast. Microbarographs thousands of kilometers away registered a sudden change in atmospheric pressure as the shock wave from the blast passed, and for several days white nights were recorded as far away as Scotland, allowing newspapers to be read at midnight, as dust thrown high into the atmosphere by the explosion was continuously illuminated by the Sun, which is not far below the horizon during the short summer nights in the north. Ice samples from Greenland also registered a thin layer of dust, thrown up by the explosion, which had slowly settled out of the atmosphere.

Leonid Kulik led the first expedition to the region in 1927. Eyewitnesses told him that people as far as 60 kilometers away had been knocked over by the explosion. To his surprise, he found no crater in the earth, but he was able to see how the forest had been flattened outward over a 30-kilometer radius from the point of impact. We still do not know exactly what it was that exploded. The absence of a crater and the pattern of felled trees suggest that the explosion was an airblast at an altitude of perhaps 8 kilometers. Almost certainly the object was fragile— a small comet or probably a carbonaceous asteroid, perhaps 60 meters in diameter, that was completely destroyed in the explosion. Though we cannot be sure, the object may have even been a fragment of Comet Encke. One thing we do know, however, is that if the explosion had happened 4 hours and 47 minutes later it would have taken place over Saint Petersburg and might have caused as many as several million deaths. Moreover, had such an explosion occurred during a period of international tension such as the cold war, it could conceivably have been taken for a sneak attack, provoking a nuclear exchange.

Small impacts such as the Tunguska event may happen as frequently as once a century. Much has been made of the fact that if the explosion was caused by a small comet, it is surprising that no one saw it as it approached the Earth. But this is no real mystery: the object came from the direction of the Sun and would not have been visible in the daytime sky. Until it enters the Earth's atmosphere, such an object is intrinsically undetectable, which means that we might very well have no forewarning of the collision.

Giordano Bruno and the Canterbury Monks

In the early 1980s, Carl Sagan hosted a well-known TV series called *Cosmos*. One episode included a fascinating sequence in which five monks from Canterbury are seen watching a fountain of flame suddenly erupt from the Moon. As Sagan explains, what these monks may have been watching was the collision of an asteroid with the Moon. If this theory is true, the impact of the asteroid, which was probably some 2 kilometers in diameter, created the crater now known as Giordano Bruno—rather aptly named for the last scientist to be burned at the stake as a heretic, in 1600, in part for defending the idea that the Earth revolves around the Sun. The episode was based on the following passage from the twelfth-century *Chronicle of Gervase of Canterbury:*

> In this year, on the Sunday before the Feast of St. John the Baptist, after sunset when the Moon had first become visible a marvelous phenomenon was witnessed by five or more men who were sitting there facing the Moon. Now there was a bright new Moon, and as usual in that phase its horns were tilted towards the east; and suddenly the upper horn split in two. From the midpoint of this division a flaming torch sprang up, spewing out, over a considerable distance, fire, hot coals, and sparks.[4]

The date of this incident is commonly given as June 18, 1178. Some people have gone so far as to suggest that this event, along with the Tunguska fireball in 1908 and a number of small impacts on the Moon that have repeatedly occurred late in the month of June, were all caused by fragments of Comet Encke, whose orbit crosses the Earth's toward the end of June every year.[5]

Gervase is remembered as a good and reliable historian, not known for the excesses of exaggeration characteristic of many of his contemporaries. Certain of the astronomical events he recorded, such as an observation of Mars passing in front of the disk of Jupiter, have been confirmed by modern calculations. But could he really have been describing a major impact on the Moon—one that took place less than a thousand years ago? Indeed, there is one apparent problem with Gervase's account that some have regarded as fatal. The issue turns on whether a reference in the Latin to the "Lord's Day" can acceptably be translated as "Sunday." We know that there was a new moon on Saturday, June 17, 1178. On the evening of the following day, June 18, the crescent moon would have been very small and very low on the horizon, so much so that it would have been all but invisible in the evening twilight, making it highly unlikely that anyone could have observed it. In this case the information Gervase provides must be in error. But as the British astronomer Graeme Waddington has noted, the full text of Gervase's account actually mentions two dates, and the second, Monday, June 19, presents less of a problem. By then the Moon would have been a somewhat thicker crescent, higher in the sky after sunset and thus more readily visible.

The crescent moon is important to the Islamic faith. Because the start of Ramadan is determined by the first sighting of the crescent moon in the sky after the new moon, reliable records of sightings are of considerable value, and over the centuries tables have been drawn up that list the date on which the crescent moon first became visible at different points on the Earth. According to these tables, in the Canterbury area at a quarter to nine in the evening on Monday, June 19, 1178, the crescent moon would have been low in the western sky—only 2 degrees above the horizon—and would have set an hour and twelve minutes after the Sun.[6] Thus it should have been visible from Canterbury, although it might not have been an easy object to spot unless it were set against a clear horizon. Apart from that, though, there is no good reason to suppose that in those days—long before urban sprawl, smog, and light pollution robbed man of the chance to see the stars— people never took the time to watch the thin crescent moon slide slowly beneath the horizon when they spied it for the first time in the evening sky after a new moon.

There are other difficulties, however, one of which has to do with the observation that the Moon's upper horn was split in two. The incident recorded by Gervase is far from the only account of a split moon in contemporary chronicles. In fact, the image of a double or split moon was commonly used to describe eclipses, which were often understood as a dark moon covering up the bright one. At the same time, "a flaming torch sprang up, spewing out, over a considerable distance, fire, hot coals, and sparks" does sound like a description of the fountain created by a great explosion, and the position indicated in Gervase's text—the "upper horn" of the Moon, or the northeast of the disk—corresponds well to that of the crater Giordano Bruno, situated just over the edge of the visible face of the Moon in the northeast. Skeptics, however, suggest that Gervase was simply describing an atmospheric effect produced by the Moon being so low on the horizon. Despite having watched many moonrises from the observatory on Tenerife I have yet to observe this effect firsthand, but I have been reliably informed that a rising or setting Moon can indeed appear to split in two as a result of atmospheric refraction. Furthermore, while Giordano Bruno is obviously a young crater—it is surrounded by a characteristic pattern of bright rays created by the "splash" of material fresh from the impact—it is not necessarily all that young. Although some have argued that other bright-ray craters like Copernicus and Tycho are only 50 to 100 million years old, the Surveyor 3 probe has now given us a fairly reliable estimate that Copernicus is about 810 million years old. Tycho is almost certainly much younger, but clearly neither of them is under a thousand years old. What is more, at this point in the evolution of the solar system such a major impact should occur on the Moon only once every several million years. A billion years ago such collisions were still quite common, but now that the solar system has largely been swept clean of debris they have become exceedingly rare.

But perhaps the most serious objection to the theory that the event described in Gervase's text was in fact a gigantic lunar impact is one made in 2001 by a young graduate student at the University of Arizona's Lunar and Planetary Laboratory, Paul Withers. As Withers pointed out, such a massive impact on the Moon would have flung millions of tonnes of material into space, part of which would subsequently have fallen on the Earth. This would have given rise to an enormous shower of meteors, more correctly called a "meteor storm," that could easily have lasted a full week. No such shower was observed from anywhere in the world, however, which casts grave doubt on the possibility that Gervase actually witnessed the impact of a large asteroid on the Moon.

Did Gervase record the formation of Giordano Bruno? We just don't know. The balance of probability says no, but it is not altogether impossible that he did.

Triumph or Disaster?

Mass Extinctions

Most reasonably well-educated people are familiar with the extinction of the dinosaurs. Although some of the details remain obscure, there is no question that around 65 million years ago not only the dinosaurs but an important fraction of all life-forms on Earth suddenly disappeared. What is less well known is that this was just one of a number of similar events that have taken place over the last 500 million years. And it was not even the largest of them. Around 220 million years ago nearly half of all species of animals vanished, clearing the Earth for the arrival of the dinosaurs. Prior to that, some 245 million years ago, an even bigger wave of extinctions took place—the "Great Dying," as it has been termed. Some estimates suggest that as much as 96 percent of all species disappeared at that time. Nor was the death of the dinosaurs the end of it. Another massive extinction occurred around 35 million years ago, which killed off nearly as many species of animals as was the case when the dinosaurs vanished. What caused these massive extinctions of life on Earth?

Just as we know that the Moon has been hit by some very large objects over the past few hundred million years, we know that the Earth has too. Over recent decades giant impact craters have been discovered in increasing numbers, and most experts now believe that the extinction of the dinosaurs was caused, at least indirectly, by the impact of a very large asteroid on the Yucatán peninsula. The Chicxulub crater is 180 kilometers in diameter, but it lay unrecognized for many years because it is largely hidden on the sea floor and can most readily be detected by the way that it distorts the local field of gravity. Along with its enormous size, the age of the crater—65 million years, which is exactly how long ago the great extinction of the dinosaurs occurred—is a smoking gun.

Even before this crater was located, however, scientists strongly suspected that some sort of catastrophic collision was responsible for the death of the dinosaurs. All over the world, a thin layer of clay rich in iridium separates the rocks younger than 65 million years old, which preserve no trace of the dinosaurs, from the fossil-rich rocks older than 65 million years. Iridium is a exceedingly rare element on Earth, but it is much less rare in meteors. The existence of the iridium layer in places as far apart as Italy and New Zealand points to a huge impact that distributed debris across the globe. By itself, the blast from the impact would not have caused the extinction, but the change in weather patterns, the nuclear winter, and the years of darkness would have killed plants and destroyed the food chain, thereby wiping out all but the most adaptable species. Added to this is the very real possibility of long-term climate change and, in the event of an impact in the sea, the likelihood of major alterations in ocean currents (to say nothing of giant tsunamis) that would massively disrupt the ecosystem both in the oceans and on land.

Other extinctions occurred close to the times when other huge craters are known to have formed. The large Triassic extinction, which occurred somewhere in the neighborhood of 215 to 225 million years ago, coincides with the creation of the giant Manicouagan crater in Quebec. This crater, which has a diameter of about 100 kilometers, is estimated to be between 206 and 214 million years old. The Eocene extinction 35 million years ago closely corresponds to the age of the 28-kilometer Mistastin Lake crater, which lies in Newfoundland and Labrador and is some 34 to 42 million years old, and to the age of the similarly sized Zaragoza crater in Spain (although the nature of this crater has yet to be firmly established, and not everyone agrees that it was produced by an impact). In addition, the Great Dying was recently linked to fragments of meteorite found in rocks in Antarctica dating back roughly 245 million years. Even more spectacular, though, was the announcement made in May 2004 by a team of scientists led by Luann Becker from the University of California at Santa Barbara. Becker's team claims to have discovered evidence that a crater some 125 kilometers in diameter located in the seabed off the northwest coast of Australia—the Bedout Crater— was produced by the impact that caused the great extinction 245 million years ago.

Although the link between impacts and extinctions is now widely accepted, not all scientists are convinced—and the idea that mass extinctions may be periodic has proved even more controversial. In 1984, a number of different studies appeared suggesting that approximately every 30 million years a mass extinction takes place. As we have seen, in the largest of these some 96 percent of all species of life on the planet may have perished. But what would cause such periodic extinctions? Most probably the explanation is celestial. At one point a theory was proposed that the Sun has a dim red-dwarf companion, christened Nemesis, that passes through the Oort Cloud of comets around our solar system every 28 million years, sending a deadly rain of comets toward the Sun, some of which would collide with the Earth. But the orbit of such a body would not be stable, and it would only be a matter of some hundreds of millions of years before it escaped from the Sun's gravitational field. Worse yet, careful searches have failed to reveal its whereabouts. It is therefore all but certain that Nemesis does not exist.

An alternative, and more satisfying, explanation for periodic extinctions has to do with the galactic tide—the movement of our Sun in relation to the plane of the Galaxy. The Sun slowly swings up and down, moving in and out of the galactic plane (the Galaxy's "equator"), which it crosses every 30 million years. Some have therefore suggested that the collision of the solar system with the giant molecular clouds—huge, relatively dense clouds of dust and gas that lie in the plane of the Galaxy—could also lead to showers of comets. At the maximum points in this cycle, our solar system moves 230 light-years above and below the galactic plane. At the moment we are about 25 light-years above the plane and moving up, after our last crossing about 4 million years ago. There are no giant molecular clouds anywhere nearby, and so during this last crossing we probably did not encounter such a cloud. Indeed, according to the 30-million-year cycle, the most recent wave of extinctions should have happened about 5 million years ago, but evidently it did not. In any case, though, it seems we are safe from extinction for at least another 25 million years, until we next cross the plane on our way down, toward the south galactic pole.

Near-Earth Asteroids

Two types of object constitute a potential hazard for the Earth. One of these is comets. The chief worry is that a new comet—that is, one that has never previously passed close to the Sun (although it is really as old as the solar system itself, just like any other comet)—could fall in from the outermost part of the solar system and collide with the Earth. Such impacts are rare, but they are especially dangerous because they typically happen at high velocity. New comets, which can come at us from any direction, fall into the inner solar system at a rate of perhaps ten per year. Fortunately for us, no known periodic comet poses a threat, at least not over the next millennium. More immediately dangerous are the asteroids, some of which come uncomfortably close to the Earth, although when an asteroid does collide with the Earth, it generally does so at a much lower velocity than a comet. Asteroids orbit the Sun in the same direction as the Earth, and thus a collision is always the result of a tail chase and takes place at a relatively slow speed.[7]

Astronomers divide asteroids into several classes. The overwhelming majority of asteroids orbit the Sun between Mars and Jupiter and are totally inoffensive. They are not even particularly hazardous to space missions, inasmuch as no probe that has crossed the asteroid belt has ever been hit. A small percentage of known asteroids—about 0.1 percent—cross the orbit of Mars, however, and are classed as near-Earth asteroids, or NEAs. Of these, about 20 percent have orbits that might someday bring them quite close to the Earth. This last group includes the "potentially hazardous asteroids," or PHAs. To summarize just a few facts about asteroids:

- As of February 27, 2005, there are 3,199 NEAs. More than half of these (1,704) have been discovered since 1997 by the U.S. Air Force's LINEAR Telescope in New Mexico.
- A PHA is an asteroid that has a diameter of at least 300 meters and that at its closest approach will be less than 7.5 million kilometers from the Earth— about seventeen times the distance of the Moon. To date, 628 PHAs have been discovered, 329 of them by LINEAR and 70 by various surveys based at Mount Palomar in California. The total number is unknown, but calculations based on the number of known objects suggest that there could be as many as 1,200 PHAs whose diameter is at least 1 kilometer, and many more with a minimum diameter of 300 meters.
- If a stony asteroid 300 meters in diameter hit the Earth at a speed of 15 kilometers per second, its impact would be the equivalent of a 1,100-megatonne bomb (or about sixty thousand bombs the size of that dropped on Hiroshima). It would produce a crater some 3 kilometers in diameter and cause catastrophic damage, with the effects of the blast stretching across an entire continent.
- The largest known PHA is twenty times larger than such an asteroid and eight thousand times more destructive.

Close Approaches to Earth

Table 9.1 lists the twenty-six asteroids that have passed the closest to Earth to date, seventeen of which came nearer than the Moon—although many more asteroids will have passed as close as these, or closer, without being detected. Only one of these seventeen asteroids, 2002 MN, was as big as or bigger than the ob-

Triumph or Disaster?

Table 9.1 The closest near-earth asteroids

Designation	Distance from Earth (kilometers)	Diameter[a] (meters)	Date of encounter (TT)
2004 FU$_{126}$	12,900	9	2004 March 31.65[b]
2004 FH	49,400	33	2004 March 18.98
2003 SQ$_{222}$	83,800	4	2003 Sept 27.96
1994 XM$_1$	107,700	11	1994 Dec 9.79
2002 XV$_{90}$	118,200	45	2002 Dec 11.35
2002 MN	119,700	94	2002 June 14.09
1993 KA$_2$	148,100	7	1993 May 20.86
2003 XJ$_7$	148,100	33	2003 Dec 6.79
2003 SW$_{130}$	161,600	7	2003 Sept 19.24
2004 OD$_4$	166,300	21	2004 July 16.81
1994 ES$_1$	169,100	9	1994 May 15.72
1991 BA	170,500	9	1991 Jan 18.72
2004 FY$_{15}$	239,700	26	2004 March 27.48
2004 HE	277,100	21	2004 April 18.01
2003 UM$_3$	281,300	11	2003 Oct 12.07
2001 BA$_{16}$	306,700	31	2001 Jan 15.85
2003 HW$_{10}$	330,600	16	2003 April 29.27
Moon	384,400	3,475,600	
2002 GQ	421,900	25	2002 March 31.15
1995 FF	432,300	23	1995 March 27.15
1996 JA$_1$	453,300	355	1996 May 19.69
2003 XV	457,800	21	2003 Dec 6.95
1991 VG[c]	457,800	8	1991 Dec 5.39
2002 EM$_7$	465,300	59	2002 March 8.04
2002 CB$_{26}$	480,200	19	2002 Feb 8.80
2004 DA$_{53}$	488,000	11	2004 Feb 24.30
2003 LW$_2$	495,200	26	2003 June 1.22

[a] These are estimated diameters, based on the assumption that the asteroid reflects 15 percent of the light that falls on it. A brighter asteroid—one that reflects more than 15 percent of the light—will appear to us to be larger than it actually is, and so we will overestimate its diameter (thus making the estimated figure in the table too large). Likewise, a darker asteroid will seem smaller, leading us to underestimate its diameter.

(continued)

Table 9.1 *(continued)*

ᵇPer astronomical convention, dates are given as year, month, day, and time of day, with the last expressed as a decimal fraction (such that "0.10" would be 2:40 A.M.).

ᶜThis object may not be an asteroid at all but instead may be the third stage of one of the Apollo moon rockets. Most of these rockets were deliberately crashed into the Moon, but a few went into solar orbit and will consequently approach the Earth from time to time. This identification of 1991 VG has yet to be confirmed, however. In the meanwhile, an apparently serious suggestion has been made that the object could be an alien reconnaissance probe.

ject that exploded over Tunguska, and only one object in the entire list, 1996 JA₁, qualifies as a PHA, that is, an object more than 300 meters in diameter that could cause continentwide destruction. A hypothetical impact with 1996 JA₁ would have an explosive force of around 1.5 *million* megatonnes. However, many of these objects are at most 10 meters across (which, given that the word usually refers to objects several kilometers across, perhaps makes the term *asteroid* a little misleading), with masses ranging from 1,000 to 5,000 tonnes. None of these small objects would cause truly cataclysmic destruction unless they failed to break up in the atmosphere and then struck the ground in a densely inhabited region. That said, even a 1,000-tonne asteroid—the smallest of those that have passed nearby us— packs a considerable punch: such an object would strike the Earth with the force of a small nuclear warhead.

Let us look at an example of what could happen. On August 10, 1972, many people were stunned to see a brilliant fireball streak across the skies of Oregon in broad daylight. Visible for two minutes, the fireball was in fact a 9,000-tonne rock that flew through the atmosphere at 14 kilometers per second at a height of only 50 kilometers before escaping into space. A 9,000-tonne rock is really quite small—under 20 meters in diameter—but had it hit the ground the explosive force would have been some 200 kilotonnes—roughly ten times the force of the bomb that destroyed Hiroshima. Fortunately, most rocks of this size would explode in the atmosphere, just as the Tunguska object did, rather than smashing into the ground, which would lessen (although perhaps not eliminate) the damage they could do. Only rocks that contain a lot of iron would be sturdy enough to survive the shock of the encounter with the denser layers of Earth's atmosphere, which would otherwise cause them to break up much as if they had been thrown against a brick wall. Once again we see that the Earth's atmosphere offers us excellent protection. Above a certain size, however—some tens of meters in diameter—even an asteroid that contains relatively little iron will punch through the Earth's at-

Triumph or Disaster?

mosphere as if it simply were not there. In this case, the only effective defense is not to be anywhere nearby when the asteroid hits the ground.

One interesting object that was once high on the list of asteroids to have made the closest approaches to Earth, although it now ranks only forty-fourth, is (69230) Hermes. The asteroid was observed for just four days in 1937 as it flew past the Earth at less than twice the distance of the Moon. "FIVE HOURS SAVED WORLD FROM DISASTER" screamed the headline in the *London Daily Mirror* on January 10, 1938, when the news was released. As the story went on to say:

> Five and a half hours saved the world from annihilation . . . . Horror-stricken astronomers in different parts of the world saw the planet crashing straight towards the earth. They watched it until it was within 400,000 miles of the earth's crust. Then it veered off and rushed by.

Apart from proving that a sensationalist press existed even in the 1930s, those few lines manage to contain four major scientific inaccuracies: Hermes is not a planet, it was not heading straight for the Earth, it did not "veer off" but simply remained in its own orbit at a safe distance from us, and it was not massive enough to have annihilated the Earth. Because it was observed for such a short time in 1937, the orbit that could be calculated for it was only approximate, and Hermes disappeared again as swiftly as it had appeared.

Hermes was lost for sixty-six years before it was relocated by astronomers at the Lowell Observatory, in Arizona, in the autumn of 2003, at which point a definitive orbit could finally be calculated. Moreover, radar observations made by the huge radio telescope at Arecibo, in Puerto Rico, have now revealed that Hermes is not, in fact, a single object. Rather, it is two almost equally sized objects that are so close to each other that they almost touch and that orbit around each other every twenty-one days. In view of its size—Hermes is approximately 0.9 kilometers in diameter—and the fact that it might make a very close approach to our planet, and given that its whereabouts were for so long a mystery, scientists frequently cited Hermes as a potential doomsday asteroid. But this is to exaggerate the danger. Although it is very far from tiny, Hermes is actually too small to threaten us with annihilation—much smaller than the object that put an end to the dinosaurs' reign on Earth, which was probably some 20 kilometers across. What's more, studies of its orbit have shown that, at least for the time being,

Hermes is not going to come very near the Earth—not even as close as the Moon. In other words, Hermes does not pose a threat to the survival of the human race. Given a slightly different orbit, however, an asteroid like Hermes could conceivably kill at least tens if not hundreds of millions of people. It would also make conditions on our planet extremely inhospitable for quite some time to come, even if humanity as a whole would ultimately survive the disaster.

As we have seen, the Earth is a tiny target in space, and the odds are millions to one against a collision with an object of any significant size. All the same, a wise person buys insurance, not because they expect that their house will burn down, but because they wish to guard against the remote chance that it will. In the present case, taking out insurance consists of detecting potentially dangerous asteroids well in advance and learning how to protect ourselves from them. Once a PHA is identified, a very careful investigation of its orbit is carried out. Sometimes we are lucky enough to discover that on some earlier occasion the asteroid was photographed when it passed close to the Earth and that the images have been lying unrecognized in an archive since then. When an asteroid has been observed on various occasions separated by at least a few years, its orbit can be calculated with great precision, and its position can thus be predicted centuries in advance. But even when an asteroid is newly discovered and our information about it is scant, we can still calculate possible close approaches to the Earth several decades into the future.

How can this be done when we have only a few observations of a new object? The explanation has to do with a radical change in technology. To calculate the orbit of an asteroid or a comet we need to obtain precise measurements of its position in the sky at different times. Twenty years ago, most measurements were made photographically and without the benefit of computers, and, because the work was slow and difficult, few observatories specialized in collecting such data. Now, however, we have CCD cameras equipped with sophisticated computer programs that are cheap enough and simple enough to use that many amateur astronomers are able to make measurements as accurate as, or even better than, those that the best professional observatories could produce twenty years ago. Many amateur astronomers now make an invaluable contribution to the study of asteroids by dedicating themselves to observing and measuring precise positions for the objects flagged in different Internet databases as PHAs.

As a result, not only is more data available more rapidly, but also the quality of the data is much higher. In just a few days, we are generally able to make a fairly accurate approximation of the orbit of a new object, usually one sufficient to indicate roughly where it will be twenty or thirty years from now, from which we can judge whether the object is potentially hazardous. In most cases, calculating the orbit reveals that the object is of no great interest, but every now and then the calculations deliver a warning, for example, that an asteroid is going to pass close by the Earth at some future date. Such a warning triggers closer study of the object, which allows its orbit to be far better defined. Thus we know that over the next two centuries no *known* asteroid will pose a danger to the Earth. According to current predictions, the closest approach will be made by 2004 MN_4, an object perhaps some 600 meters in diameter, at 12:50 UT on April 13, 2029, when it will be approximately 33,000 kilometers from the Earth—a reassuringly safe distance. Moreover, the orbit of this asteroid is at this point pretty firmly established, so these numbers are unlikely to change much as new observations are made. For a time, calculations suggested that asteroid 2000 WO_{107}, which has a diameter of roughly 550 meters, would also pass by the Earth at a distance of only 81,000 kilometers on December 1, 2140. But in this case continued study of the orbit revealed that the asteroid will actually be about four times further away than originally predicted.

1997 XF_{11}, 1950 DA, and Others

Impact scares are actually nothing new. For centuries comets have been regarded as a potential source of catastrophe. Donald Yeomans, of NASA's famed Jet Propulsion Laboratory, recounts the story of an end-of-the-world scare that took place in Paris in 1857, when it was predicted that a large comet would strike the Earth on March 13. Tens, perhaps hundreds, of thousands of people genuinely believed that the apocalypse was nigh. Then, in 1910, the news that the Earth would pass through the tail of Comet Halley caused a panic that resulted in all manner of scams, including the sale of pills, potions, and charms, as well as gas masks to protect people from toxic vapors in the comet's tail. Sales of oxygen cylinders rocketed as banner headlines proclaimed, "Comet to Bounce off Earth Today—Tough Old World to Survive."[8]

What changed our perception of the threat of possible collisions, however, was the saga of the asteroid 1997 XF_{11}, which started with the discovery by Jim

Scotti and the Spacewatch telescope, at Kitt Peak in Arizona, of an asteroid in the constellation of Cancer. The object had a magnitude of about 19 and appeared in three exposures made on December 6, at 11:20, 11:50, and 12:22 UT. The story nearly ended there because bad weather stopped Scotti from being able to follow up on his discovery. Had it not been for the determination of two Japanese amateur astronomers, the asteroid would probably have been lost then and there. As it was, though, only three days later, on December 9, Atsushi Sugie recovered 1997 XF$_{11}$ with his 60-centimeter telescope and made six measurements of its position. He was able to locate it again on December 18 and made four more measurements, which ensured that the asteroid would not be lost again. On December 21 a second Japanese amateur, Takuo Kojima, also observed 1997 XF$_{11}$ with his 25-centimeter telescope and made six additional precise measurements of its position at different times. These observations were enough to tell us that the asteroid was worth keeping an eye on. Not only was it rather large, about 1.8 kilometers in diameter, but the minimum distance between its orbit and the Earth's was unusually small. Fortunately, other observers made yet further measurements when they realized that 1997 XF$_{11}$ was potentially hazardous.

By February 1998 astronomers had determined that in October 2028 the asteroid would apparently pass by us at about twice the distance of the Moon. No other asteroid of such size would come so close to the Earth within the next century, but with less than three months' worth of data a great deal of uncertainty remained about its orbit. More data were needed. On March 3 and 4, the asteroid was observed with the 76-centimeter telescope at the McDonald Observatory, located in the Davis Mountains some 450 miles west of Austin, Texas. These new observations reduced the predicted distance of the approach in 2028 to 48,000 kilometers, or only four times the diameter of the Earth—alarmingly close. Four astronomers calculated the orbit and reached similar conclusions. In fact, one suggested that the asteroid would come closer still. The prospect of such a close approach pointed to a new potential hazard—that the Earth's pull of gravity could conceivably change an asteroid's orbit, putting it on a collision course with Earth. Were a collision to occur with such a large asteroid, the explosion would be the equivalent of perhaps a hundred times the entire nuclear arsenal of the world at the time of the cold war.

At this point an official announcement was made by the IAU's Minor Planet Center and the Jet Propulsion Laboratory calling attention to the close approach

Triumph or Disaster?

in 2028, as well as to the somewhat lesser danger of an impact around 2036. Thanks to the announcement, within twelve hours traces of the asteroid had been located in photographs taken in March 1990 with the 46-centimeter Schmidt telescope at Mount Palomar Observatory, outside San Diego. Curiously, the old photographic plate had been intended precisely to detect asteroids that might pass near the Earth, but the faint trail of 1997 XF_{II} had been missed. The asteroid was also missed earlier in 1997, when it passed relatively close to the Earth and was three magnitudes brighter than it was when the Spacewatch telescope finally spotted it later in the year. The older observations, technically known as a "precovery," enabled the orbit of 1997 XF_{II} to be calculated over a total of seven years—more than four of its orbits around the Sun.[9]

These new calculations yielded two main results, both of them comforting. Most immediately, they demonstrated that in 2028 the asteroid would pass the Earth at a distance of 960,000 kilometers—more than twice the distance of the Moon and twenty times further away than the 48,000 kilometers previously predicted. This may sound like a fantastically large error on the part of the astronomers, but in fact it meant that in calculating the asteroid's orbit more than thirty years into the future, the position they had predicted was in error by just five and a half hours along its orbit—an astonishingly tiny amount. At the same time, the new and better calculations showed that even though between 2035 and 2037 the orbital path of 1997 XF_{II} and that of the Earth will overlap, a collision is impossible because never will the asteroid and the Earth itself be at precisely the same point at the same time. Starting in 2036 the orbits will begin to separate again, at which point the danger of a potential impact disappears entirely. In the end, though, what saves the day is only the lucky chance that the Earth will be elsewhere when 1997 XF_{II} crosses its orbit between 2035 and 2037. Were the timing slightly different, the projected ending might not be so happy.

In 2002, the prestigious journal *Science* published an article detailing the study of an asteroid known as (29075) 1950 DA. The authors calculated that there could be as much as a 1-in-300 chance that this asteroid would collide with the Earth on March 16, 2880.[10] Their announcement came on the heels of a series of minor scares of the sort that have taken place over the past several decades. On a few occasions these scares have had at least some basis in fact, although even in these cases the odds have still been overwhelmingly against the possibility of a collision. Others, such as the 1968 scare that the close pass of Icarus to Earth would cause the San Andreas Fault to let loose and San Francisco to fall into the sea, have been completely imaginary. Asteroid 1950 DA is unusual in two ways, however. In the first place, the probability of impact is by far the highest ever detected. Granted, no sensible person would wager much money on odds of 300 to 1, but when so much is at stake these odds become somewhat more meaningful, not to mention alarming. With a diameter of 1.1 kilometers, if this asteroid were to collide with the Earth, it would wreak utter devastation over a radius of hundreds of kilometers and leave behind a crater perhaps 11 kilometers wide and 1 kilometer deep. At the same time, in contrast to cases such as that of 1997 XF$_{11}$, the threat of an impact is nearly nine centuries in the future and thus lacks any sense of immediacy. Even the most farsighted of us attach relatively little importance to the fate of our descendants beyond our grandchildren—and the fate of the thirtieth generation of our descendants lies such a long way in the future as to be almost beyond comprehension.

It will be many years before we can refine our calculations of the orbit of 1950 DA sufficiently to confirm or, more likely, rule out an impact in 2880. Fortunately, by that time we will almost certainly have worked out a method for changing the course of asteroids that would otherwise come too close to our planet for comfort. In many ways the technology is already well within our reach, although in order actually to intercept an asteroid, we would probably first need to develop a manned rocket—the simplest and least expensive option would essentially be an upgraded version of the Saturn V—capable of lifting at least two or three astronauts and a sizeable payload into solar orbit. As matters stand, we also lack the infrastructure necessary to such an enormous and complicated undertaking. Then again, the most effective approaches to deflecting an asteroid from its course do not involve destroying it. They depend instead on discovering the asteroid many years before an impact could occur and then slowly but steadily altering its path— clearly not something we could accomplish overnight, or even within a single decade. Provided we have sufficient warning, though, a number of fairly simple solutions are potentially available. These include firing a powerful laser at a threatening asteroid so that the vaporized rock will act as a small rocket, giving the asteroid a gentle nudge in another direction. A deflection of a just few meters would make a difference of thousands of kilometers in the future path of the asteroid, enough to change a hit into a completely harmless near miss.

Other solutions rely on the gentle pressure of light acting over many years.

Triumph or Disaster?

The light from the Sun exerts a tiny, but measurable, pressure on objects in space. Simply put, a beam of light possesses a small amount of momentum, and when light hits an object and is absorbed, the light's momentum passes to the object, giving it a tiny push. If instead the light is reflected off the object, then twice the amount of momentum passes to it, producing twice the amount of push—still not very much, but over many years the cumulative effect becomes significant. One way of subtly but effectively altering the orbit of an asteroid would therefore be to paint one side of it white so that it reflects sunlight more efficiently. There would be no need to send in Bruce Willis with a paintbrush and a very large can of paint: the operation could be carried out quite nicely simply by spraying chalk dust on the asteroid from a respectful distance. The increased reflectivity would then translate into an additional gentle push from the pressure of sunlight. Making one side of an asteroid rougher (using conventional explosives, for example) would also cause it to absorb sunlight better, generating a similar, although smaller, effect.

Further research into how to deal with the threat posed by asteroids or comets is currently underway. Various space probes have flown close by asteroids and have photographed them as they passed, but far more was learned when NASA's probe NEAR observed the asteroid Eros, first from orbit and then by landing delicately on its surface. The probe gave astronomers their first detailed look at the structure and composition of an asteroid that could one day—that is, in perhaps a hundred million years—strike the Earth. Similarly, on July 5, 2005, when the Deep Impact probe smashes a small impact probe—a 350-kilogram copper cylinder—into the nucleus of Comet Tempel I, we will learn an enormous amount about comets, including their internal structure. The Deep Impact probe should provide us with more reliable information than we currently have about whether comets are reasonably solid objects that, while they would be difficult to stop, could be redirected or are instead extremely fragile objects that fall apart relatively easily, making it a simple matter to destroy them. Such information will help us to plan how we might best defend the Earth against a potential collision with a comet.

Hunting NEAs

Many observatories throughout the world now work to discover NEAs, and especially PHAs, and then to ensure that the objects discovered are observed thoroughly enough to allow their orbits to be calculated with great precision. A fair number of these telescopes are fully automated and employ highly sophisticated technology, such as the LINEAR telescope, which in 2001 alone made 2,910,824 observations of 124,644 asteroids and 81 comets. In that same year over three hundred observatories around the globe reported a total of almost four and a half million observations, tracking newly discovered asteroids and gathering better information about the orbits of known ones. As of February 24, 2005, the Minor Planet Center's database lists a staggering 29,065,968 observations of 267,951 objects. At this point, few asteroids that pass close to the Earth in the northern hemisphere are overlooked, although the coverage of the southern hemisphere is far more patchy and until recently was almost entirely left to amateur astronomers.

Even so, occasionally an asteroid is missed. Sometimes this happens because, as the asteroid approaches the Earth, it is traveling away from the Sun and appears only during daylight hours, with the result that it cannot be seen until it has already passed by us and into the nighttime sky. Nothing can be done about such asteroids. On March 31, 1989, the 350-meter-diameter asteroid (4581) Asclepius was discovered by astronomers at the Mount Palomar Observatory. Once its orbit was calculated, it became clear that eight days earlier, coming out of the Sun, it had passed by the Earth unobserved at little more than the distance of the Moon. On other occasions, owing to bad luck, an asteroid simply may be in the wrong place at the wrong time and therefore be overlooked by automated search programs.

On March 2, 2002, for instance, an amateur astronomer in Spain, Rafael Ferrando, was observing the night sky with his 30-centimeter telescope and its CCD camera. After measuring the position of three comets and a supernova he pointed his telescope at asteroid 2000 QW$_{65}$, discovered a year and a half earlier, so that he could also measure its position and thus define its orbit better. To his amazement he noticed an extremely faint trail in the upper part of the image. Several more exposures revealed that it was apparently a nearby asteroid moving quickly across the field of view of his telescope. The Minor Planet Center subsequently designated the object 2002 EA and, after examining observations from other observatories, confirmed that it was indeed an NEA—only the fifth to be discovered by an amateur astronomer.[11] Less than two weeks later, on March 15, asteroid 2002 EA passed by the Earth at a distance of 8.5 million kilometers—which, as subsequent investigations into its orbit established, was just about its nearest possi-

Triumph or Disaster?

ble approach to Earth. Because it will never come within 7.5 million kilometers of our planet, 2002 EA does not qualify as a PHA. All the same, it is hardly reassuring that it took an amateur astronomer with a backyard telescope to locate it. The case of 2002 EA, and likewise that of the 9,000-tonne rock that unexpectedly streaked across the Oregon sky, should serve to remind us of an important principle: if you know about it, then it is probably not dangerous. As we have seen, so far none of the comets and asteroids that we have identified and tracked pose any significant threat to the Earth. It's the unknown ones, lurking somewhere out there in space, that should worry us.

Homo Sapiens: The New Dinosaurs?

Could the human race disappear one day just as the dinosaurs did, wiped out by the impact of a giant comet or asteroid? Yes, it is possible, but it is not very likely. No known asteroid in the PHA class is as large as the one that fell on the Yucatán some 65 million years ago, and it would be astonishing if our observations thus far had missed a near-Earth asteroid of such colossal proportions. Besides, if such an object does exist, the Earth is still only a tiny target. Sometime within the next several decades, moreover, we are likely to have developed the means to defend the Earth successfully against a strike by an asteroid or comet, provided we have sufficient lead time. Chances are, then, that as a species we are safe until the solar system next crosses the plane of the Galaxy, in another 25 million years— and if we survive that long, it is doubtful we will do so as the same human beings that exist today.

Setting aside mass extinctions, though, how real is the threat that an asteroid will collide with the Earth within the next few centuries, or that you or I will be killed in such a collision? As we know, in the shorter term the probability that our planet will be struck by an asteroid of any considerable size is minuscule in the extreme. In fact, the chances of an event even on the relatively manageable scale of the explosion over Tunguska are still exceedingly small. As I write, in August 2004, the cumulative probability—that is, the sum of all the individual probabilities—that one of the asteroids being tracked by the Jet Propulsion Laboratory's Sentry System will collide with the Earth can run as high as 1 in 300, which certainly seems a frightening statistic. Luckily, most of these asteroids are small objects—no more than perhaps 30 meters in diameter, and generally much smaller—that would in all likelihood explode in the atmosphere. They would still have the force of a several-megatonne bomb, however, and *could* thus do serious damage in the area where they fell.

Obviously, were a large asteroid to collide with the Earth, the results would be catastrophic. But such an event is incredibly rare—so rare that, according to current calculations, the likelihood that any particular individual will die as the result of an asteroid impact is about 1 in 20,000—roughly the same as the odds of dying in an airplane crash. At the same time, we do not necessarily laugh at people who are afraid of flying on account of what might seem to some a negligible risk. Or, as astronomer Michael Paine—a well-known authority on asteroid impacts—recently pointed out, there is, in any given year, a 1-in-6,000 chance that an asteroid large enough to kill at least a million people will strike the Earth. Such a highly improbable event may hardly seem like something to be concerned about. And yet, as Paine rightly notes, if we were talking about the danger posed by some sort of man-made creation—the chances of a nuclear accident, for example—it is unlikely that we would find such odds acceptable.[12] Let us also not forget that a very good-sized asteroid, (29075) 1950 DA, with a diameter of about 1.1 kilometers, is out there, and we have not yet been able to rule out the possibility that it could strike the Earth on March 16, 2880. Granted, there is roughly a 99.7 percent probability that it will miss us. But when an object is capable of blasting a hole in our planet at least 10 kilometers wide, with the impact and its aftereffects potentially killing many millions of people, prudence suggests that we accord it the greatest respect, no matter how far the odds appear to be in our favor.

Even if it were only one of the smaller asteroids that hit the Earth, the impact would be devastating, at least over a localized area—equivalent to the destruction caused by a large nuclear bomb. A land strike in a densely populated area of Europe or Asia would obviously take a heavy toll in lives. In addition, the huge quantities of dust thrown up into the atmosphere by the impact *could* cause a nuclear winter, with the danger of subzero temperatures, over a large part of the planet, a situation that could last for years or even decades. If a collision were to occur, however, it would most likely be a sea strike, for the simple reason that the Earth's surface consists more of water than land. But this is no comfort, for a sea strike is potentially the most destructive type of impact. A major strike in the North Atlantic would, for example, produce tidal waves that could severely dam-

Triumph or Disaster?

age or even completely destroy coastal cities such as New York, Washington, Los Angeles, Lisbon, Oslo, Copenhagen, or Miami. Even a relatively small asteroid would slice through the water and crash into the ocean floor, hurling massive amounts of mud into the atmosphere, which would have the same effect as dust, blocking out the sun's light and again possibly creating a nuclear winter. Such a collision would also send billions of tonnes of water into the atmosphere, resulting in severe storms, including hurricanes, for weeks following the impact. A sea strike might also change ocean currents and therefore bring about significant changes in climate—possibly eliminating the so-called Gulf Stream, for example, which moderates the climate of Europe. The human race would itself survive, but the death toll could easily be in the tens of millions, and the costs would be appalling in other ways as well.

Where, then, does that leave us? The next time someone announces that there is a 1-in-500,000 chance that asteroid (2007) BF will strike the Earth on June 5, 2019, should we panic? The answer is emphatically "no." As often as not, such alerts are based on very minimal data that, despite being so inadequate, have been extrapolated decades into the future—and in any case the fact remains that no known asteroid poses an immediate threat to the Earth. Granted, there are plenty of unknown ones that could conceivably strike us (or, more likely, explode in the atmosphere over our heads) and potentially wreak terrible destruction over a specific area. The probability that you or I would happen to be in that area at the time is, however, extremely minute.

In short, it is pointless to become hysterical over an infinitesimal risk from a known object whose path we can predict. There is, however, a big difference between "infinitesimal" and "zero." After all, if you keep on playing the lottery year in and year out, sooner or later your number will come up. The odds are extremely long for any single draw, but someone always wins. In the same way, the Earth *will* someday suffer a major asteroid impact—not tomorrow, and probably not this century, but eventually. If it is foolish to panic over a very tiny possibility, to fail to be concerned about a serious long-term risk posed by a large category of objects is equally foolish. No matter how remote the threat may be today, we cannot afford to ignore the danger and take no steps to protect our planet. If we do, then inevitably, one day in the future, we will find ourselves on the losing end of the wager.

SUGGESTIONS FOR FURTHER READING

Popular Books

Larry Niven and Jerry Pournelle, *Lucifer's Hammer* (London: Orbit Books, 1977).

> *A classic science fiction novel that describes what the impact of a large comet on the Earth could do to human civilization. Well researched and full of solid scientific information, even if in some places the science is slightly out of line with what is now mainstream astronomical opinion. All the same, a novel that continues to offer a highly plausible "what if" scenario.*

Carl Sagan, *Cosmos* (New York: Random House, 1980).

> *A book based on Sagan's popular television series. One chapter is devoted to possible collisions and other such catastrophes and includes Sagan's discussion of whether Gervase of Canterbury recorded the formation of the lunar crater Giordano Bruno.*

Walter Alvarez, *T. Rex and the Crater of Doom* (Princeton: Princeton University Press, 1997).

> *A true-life detective story about how Alvarez discovered the details of the impact that led to the extinction of the dinosaurs. An excellent source of information on impacts, craters, and mass extinctions.*

Kenneth Hsü, *The Great Dying: Cosmic Catastrophe, Dinosaurs, and the Theory of Evolution* (New York: Harcourt Brace Jovanovich, 1986).

> *A fine book about how impacts and cosmic catastrophes influence evolution, as well as a useful look at how scientists uncover evidence concerning such events.*

Jack Stoneley, *Tunguska: Cauldron of Hell* (London: W. H. Allen & Co., 1977).

> *Another fascinating account of how scientists track down evidence, in this case to explain the mysterious explosion in Tunguska in 1908.*

More Advanced Reading

J. Kelly Beatty and Andrew Chaikin, eds., *The New Solar System*, 3rd ed. (Cambridge, Mass.: Sky Publishing, 1990).

> *A valuable collection that includes essays by acknowledged experts such as William Hartmann and Gene and Carolyn Shoemaker on asteroids, meteorites, and other small bodies, and on impacts. Beautifully written and prepared, and easily accessible to the advanced amateur astronomer.*

208   *Triumph or Disaster?*

On the Internet

The Cambridge Conference Network
http://abob.libs.uga.edu/bobk/cccmenu.html

> *Benny Peiser is a sports scientist and psychologist who started an (almost) daily newsletter in February 1997, which is received by more than a thousand scientists, along with a great many of the world's most important science writers. (Arthur C. Clarke has been a regular contributor over the years, as have a number of well-known scientific journalists.) The newsletter dedicates itself to the study of neocatastrophism: the fall of civilizations, asteroid impacts, climate change, and the like. A veritable treasure trove of information on astronomical and human disasters, presented from an appropriately skeptical point of view.*

The NEO Page: International Astronomy Union, Central Bureau for Astronomical Telegrams, Minor Planet Center
http://cfa-www.harvard.edu/iau/NEO/TheNEOPage.html

> *A scientific page offering information, tables, diagrams, and graphs. A valuable resource for anyone seriously interested in finding out more about NEAs (or NEOs—"near-Earth objects").*

JPL Sentry System
http://neo.jpl.nasa.gov/risks/

> *A relatively new Web page that lists all the asteroids that could conceivably hit the Earth over the next century, with links to exhaustive information on specific NEAs that allow the user to call up data on each asteroid and view its orbit. The material is very helpfully explained, and, among other things, demonstrates just how small the current risk of an impact is.*

Barringer Meteorite Crater
www.barringercrater.com/

> *An excellent Web page about the Arizona Meteor Crater that includes a detailed presentation of the consequences of the impact and a review of scientific studies of the crater. There is even a game in which users can simulate the effects of an impact of asteroids of different sizes on the Earth.*

CHAPTER 10

Goldilocks and the Three Planets

Once upon a time there were three planets: a big planet, a medium-sized planet, and a little planet. One day an astronaut named Goldilocks visited the three planets hoping to find somewhere to live. The small planet was too cold, and all the water and a lot of the air were frozen either in the huge pole caps or in the ground. Then Goldilocks went to the medium-sized planet. But that one was far too hot, and so much gas had been released from the rocks that the atmosphere was crushingly thick. Finally, she went to the big planet, and it was just right. It was warm, but not too hot, and it had large oceans. So all in all it was just the perfect place to live on.

This is not exactly a tale worthy of the brothers Grimm, but it paraphrases a problem that has exercised astronomers for many years and that has actually come to be known as the "Goldilocks Problem." How is it that conditions on the Earth are ideal for life, whereas Venus, which is just a little smaller than Earth and a little closer to the Sun, is far too hot, and Mars, which is somewhat smaller still and only a little further than Earth from the Sun, is far too cold?

A Tale of Three Planets

To answer this question, we need to know a few basic facts about the three sister planets. As can be seen from Table 10.1, Mars is a little over half the diameter of the Earth but has barely a tenth of its mass: it is much less dense than the Earth and has a smaller iron core. Venus, which has a diameter about 95 percent that of Earth but only 81.5 percent of Earth's mass, is a tad less massive than one would expect for its size if its interior were like the Earth's, although the discrepancy is far smaller than in the case of Mars. A planet's surface gravity and escape velocity depend on its mass and diameter. On Mars, an 80-kilogram astronaut would weigh just 30 kilograms, a rather drastic weight loss of nearly two-thirds. Weight watchers would do far better on Venus, where they would be delighted to discover a definite spring in their step, their 80 kilograms having been reduced to only 70. The escape velocity, that is, the velocity that a rocket or anything else must reach in order to escape into space, also differs on the three planets. On Mars it is slightly less than half of Earth's. This makes it easy for a rocket to lift off the surface but also for the atmosphere to escape from the planet, which is why Mars has such a thin atmosphere. By the same token, because the escape velocity on Venus is also lower than that on Earth, if only by a little under 10 percent, it, too, should have a less dense atmosphere than the Earth—or so we would expect, although this is actually not the case.

The other data reveal some curious similarities and differences between the planets. Mars and the Earth have a nearly identical rotation period—a Martian day is just 41 minutes longer than our 24-hour day—and a similar degree of axial tilt. Nor are Martian seasons very different from their terrestrial equivalents. On one side of its orbital path, Mars's northern hemisphere is tilted toward the Sun and therefore gets summer, while the southern hemisphere, which is tilted away from the Sun, goes through winter. On the other side of the orbit, the situation is reversed. On Mars, however, the seasons are much longer, for the Martian year is nearly twice as long as a year on Earth—about 23 months. Some five and a half months of summer may sound like a good thing, but a winter that also lasts five and a half months is not much fun unless you happen to be a polar bear. In addition to axial tilt, orbital eccentricity—the fact that orbits are elliptical, not circular—plays a part here. People tend to assume that the Earth is furthest from the Sun in the winter. In fact, though, during winter in the northern hemisphere

Table 10.1   Earth and its neighbors

	Venus	Earth	Mars
Diameter (Earth = 1.0)	0.949 (12,104 km)	1.000 (12,756 km)	0.532 (6,787 km)
Mass (Earth = 1.0)	0.815	1.000	0.108
Surface gravity	8.60 m/sec^2	9.78 m/sec^2	3.72 m/sec^2
Escape velocity[a]	10.4 km/sec	11.2 km/sec	5.0 km/sec
Axial tilt	177.3°	23.45°	25.19°
Rotation period	243.01 days	23 hrs 56 min	24 hrs 37 min
Orbital period	224.70 days	365.26 days	686.98 days
Orbital eccentricity	0.007	0.017	0.093

[a] The escape velocity of an object is the speed at which something must be moving in order not to be pulled back toward the object by its gravitational field.

the Earth is at its closest to the Sun, and it is furthest away in the summer, although only by a difference of about 2 percent. In the southern hemisphere, it is the opposite. Thus, summers in the southern hemisphere are slightly warmer than in the north in part because the Earth is a little nearer the Sun at that time.

On Mars this orbital effect is much larger. Mars's orbit is somewhat more eccentric than Earth's, and so the difference in the amount of heat received from the Sun when the planet is at perihelion (that is, at its closest to the Sun) and at aphelion (at its furthest) is also greater. The distance of Mars from the Sun at perihelion and at aphelion differs by approximately 30 percent. Currently, in the southern hemisphere summer occurs when Mars is at perihelion and winter when it is at aphelion (although over the millennia this will slowly change as both the orbit of Mars and the tilt of its poles also change). In the northern hemisphere, however, summer takes place when the planet is furthest from the Sun and winter when it is closest. The southern summer is thus much warmer than its northern counterpart, and its winter much colder. Similarly, because Mars moves faster as it approaches perihelion, the southern summer is also much shorter than the winter. Overall, then, the southern hemisphere has a short summer that is nonetheless quite warm and a long, very cold winter, whereas in the northern hemisphere summers are cooler and much longer but winters shorter and less cold. In other words, unlike on Earth, where the two hemispheres have roughly the same climate, on Mars the climate is much more extreme in the southern hemisphere than in the northern.

All the same, Mars at least has a seasonal climate analogous to Earth's own. In contrast, Venus has a highly peculiar calendar. For one thing, it has no seasons because its poles are almost vertical. If that were not enough, its rotation period is actually about sixteen days longer than its period of revolution around the Sun. In other words, Venus has a day that is longer than its year. Moreover, its axis of rotation is "upside down." On Earth and on Mars, the Sun rises in the east and sets about 24 hours later in the west. On Venus the Sun will rise in the west and will set 59 days later in the east.[1] If astronauts ever make it to Venus, they're likely to find the Venusian calendar rather disorienting.

In short, the three planets may be siblings, but they are certainly not triplets.

Fire or Water?

Back in 1954, Isaac Asimov published a book for teenagers called *The Oceans of Venus* (originally written under the pen name Paul French). It was the third in a series of six about the adventures of a character known as the Space Ranger. Asimov pictured a world covered with deep oceans that we had colonized by building great undersea cities. At the time, the popular conception of Venus did indeed feature an abundance of water. Following a theory proposed by Fred Whipple and Donald Menzel, many scientists believed that the planet was covered with oceans, or at least that it had no shortage of water. Although the permanent layer of thick cloud prevented us from ever seeing the planet's surface, various spectroscopic measurements suggested that the clouds had a lot of water in them, making the idea of a watery world seem quite plausible. Venus was veiled, cloaked in mystery. But it seemed entirely possible that, if the clouds were lifted, the planet would be revealed as a primitive world, not unlike the Earth in the Carboniferous era, with abundant vegetation, swamps, and warm oceans.

It was not until the late 1950s that serious doubts were raised about the likelihood of oceans on Venus. As we saw, radio measurements from Earth had indicated that Venus was hot. In 1956, when the microwave radiation coming from the planet was measured, Venus turned out to be emitting this radiation at a temperature of around 330°C. Over the next few years, more accurate measurements were made both in the United States and in the Soviet Union, which raised the estimate of the planet's temperature to about 400°C—about double the temperature at which your Thanksgiving turkey is cooked. Where did this radiation come from? At the time, scientists were still heavily influenced by the view that Venus was a warm, wet planet. They thus speculated that the radiation might come from a dense, highly active ionosphere, far above the surface of the planet. Nobody could bring themselves to believe the awful, stunning truth about the planet.

In 1962 the space probe Mariner II passed by Venus and took measurements from a distance of 35,000 kilometers. Two previous probes, the Soviet Venera I, launched on February 12, 1961, and the American Mariner I, launched in July 1962, had been failures. Venera I left Earth orbit successfully and headed for Venus, but it had traveled only 7.5 million kilometers before contact was lost.[2] Mariner I did not even get that far: a minus sign had been omitted in one of the onboard computer programs, and so the rocket returned to Earth without ever leaving the atmosphere. Fortunately for the embarrassed engineers at NASA, Mariner II was a triumphant success, launched on August 27, 1962, and arriving on December 14 of the same year. The results from Mariner II were conclusive. If the microwave emissions had been coming from the atmosphere or from the ionosphere, the edge of the planet would appear much brighter than the center, whereas if the center were brighter, the emissions would have to originate on the planet's surface. Mariner II clearly demonstrated that the center of the disk was brighter. Whipple and Menzel's waterworld was dead. Even so, nobody imagined just how appalling the surface of the planet really was.

Between 1961 and 1981 a total of twenty-one missions, fifteen of them Soviet, were aimed at the planet, including four probes, two Soviet and two American, in 1978 alone. In fact, in December 1978, no fewer than seven spacecraft—five connected with the American Pioneer missions together with two Soviet probes—all landed on the planet in the space of a few days. If the Soviets had no luck at all with Mars, their success rate with Venus probes was quite astonishing: although the first three were failures, after that not a single Soviet Venus probe was lost. The Soviet Union concentrated on parachuting probes through the atmosphere. Venera 4 was the first probe to enter the atmosphere, but it ended up being crushed by the atmospheric pressure at an altitude of 29 kilometers. Venera 5 and Venera 6 managed to get only a little closer before suffering the same fate. However, Venera 7, launched in 1970, appears to have reached the surface successfully, where it survived for 23 minutes, although unfortunately its atmospheric pressure sensor failed. Venera 8 survived on the surface for 50 minutes and was able to transmit data on pressure, temperature, winds, and the level of light.

But it was with Venera 9 and Venera 10, which landed on October 21 and 25, 1975, that the Soviets' Venus program reached a triumphant climax. To the astonishment of the watching world, each probe managed to send back an image from the surface of the planet before succumbing to the truly infernal conditions.

The data from these and later probes confirmed what we essentially already knew. The surface temperature on Venus averages around 464°C—whereas the oven in my kitchen has a maximum temperature of only 250°C. The atmospheric pressure is more than 90 atmospheres on the surface (that is, ninety times the pressure on the Earth's surface), and the atmosphere is composed mainly of carbon dioxide. In fact, the atmosphere of Venus is 97 percent carbon dioxide, with the remaining 3 percent consisting almost entirely of nitrogen, along with about 0.1 percent water vapor. Add to this the fact that the clouds are made not of water but of concentrated sulfuric acid—more concentrated than in a car battery—and a picture builds up of just how appalling conditions are. Venus has a horrendous acid rain problem by any standards, thanks to which some of the earlier probes at least partially dissolved while passing through the clouds.

If Venus is not hell, it is a pretty good simulacrum of hell. The British astronomer and broadcaster, Sir Patrick Moore, aptly summed up conditions by commenting that any astronaut stepping out unprotected onto the planet's surface would be crushed, roasted, asphyxiated, poisoned, and corroded, all at once.

A Big Moon or a Little Earth?

The Mars of the space era was a huge disappointment to astronomers. Its surface is pock-marked with craters, and its atmosphere is thin: the atmospheric pressure on the planet's surface is equivalent to that at an altitude of 20 kilometers on Earth. For many years, however, scientists believed that the surface conditions on Mars were far more benign than they are now known to be. Prior to 1964, the best estimate of the atmospheric pressure was around 80 millibars, or about one-twelfth of the sea-level pressure on Earth. It was thought that the temperature could reach 20 to 30°C in the tropics in midsummer, and some experts even argued that the ground around the poles could get marshy during the spring thaw.

The results returned by Mariner IV came as a severe shock to the system. The way its radio signals dimmed as it passed behind the disk of Mars pointed to an atmospheric pressure not of 80 millibars but of a meager 4 millibars. Mariners

VI and VII not only confirmed Mariner IV's results but further indicated that the temperature of the southern pole cap was so low that it was probably not even made of ordinary ice but rather of frozen carbon dioxide—dry ice. Hopes were all but extinguished that Mars could at times thaw and, thanks to the huge amounts of polar water that would be released, become fertile. For the astronomers who studied the planet, Mars soon came to seem much more like a large version of the Moon than a small Earth.

The early Mariner probes swung the pendulum a little too far toward the negative side, however. Mariner IX revealed that the surface atmospheric pressure on Mars is in fact highly variable according to the location. Over Olympus Mons the pressure is just 2 millibars, but at deep points on the planet's surface the pressure may rise above 10 millibars. Mariner IV had, simply by chance, measured the pressure over some high ground, which accounts for its reading of 4 millibars. Mariner IX also showed that the summer temperature of the pole caps is too high for them to be made purely of solid carbon dioxide. In addition, the probe discovered craters that looked like splashes in the ground. The most logical explanation for these is that an asteroid collided with a layer of permafrost, which melted and caused a tidal wave of liquid mud to flow out from the zone of impact until it eventually dried out. Combined with dry gorges that looked like river valleys and weather systems that in some cases looked positively terrestrial, Mars began to look much more inviting.

The Viking program provided its own surprises. The Viking probes confirmed that many regions of the planet resemble massive flood plains, where huge bodies of water have scoured the landscape. Wherever the water is now, in the past there was enough of it to shape the terrain. Perhaps the greatest surprise lay in the pole caps, however. Contrary to the belief that had prevailed for some two hundred years, these are not thin caps of frost but are rather hundreds of meters thick, with a thin layer of frozen carbon dioxide on the top. There is also extensive layering, which shows the caps have expanded and withdrawn many times in the past. Craters near the edge of the poles are filled to the brim with ice, indicating that a huge amount of water is locked up in the poles—water that, if spread over the planet's surface, would create a layer not a fraction of a millimeter thick, as had been thought, but meters deep. Strange though it may seem, for some scientists this discovery was actually a crushing disappointment. They had hoped that Mars would prove to have huge quantities of carbon dioxide frozen in the poles that might be released during relatively warm epochs, giving the planet a much denser atmosphere.

As the two Viking landers and, later, the Mars Pathfinder demonstrated, however, the surface temperature on Mars never even threatens to reach freezing point. Although none of these probes landed on the equator, where the conditions could be a little warmer, the maximum temperature registered by the Vikings was around −20°C during the day, with a low point around −100°C at night, although Mars Pathfinder did at one point register a temperature as high as −4°C. The Viking probes also offered another small clue about Mars, namely, that the atmospheric pressure actually changes significantly between winter and summer. In the southern summer the south pole points toward the Sun and heats up substantially. The pole cap melts and liberates water vapor and carbon dioxide, which are absorbed into the atmosphere, raising the atmospheric pressure. At the same time, though, the north pole is cooling. As it does so, water vapor and carbon dioxide freeze out of the atmosphere and are locked in, lowering the pressure again in a constant dance. The atmospheric pressure at a given point may thus rise and fall by as much as 25 to 30 percent over a Martian year, as a similar fraction of the atmosphere freezes into the pole caps.

Why Does the Earth Not Freeze?

So we have three planets. One, Venus, is 28 percent closer to the Sun than the Earth, receives 64 percent more solar radiation, and has an average surface temperature of about 430°C. Another, Mars, is 52 percent further from the Sun than the Earth, receives 43 percent of the solar radiation that the Earth receives, and has a surface temperature of around −55°C. The average temperature on Earth is a balmy 15°C. Why such an enormous range in temperature when the amount of heat the planets receive from the Sun is not all that different?

On Mars, the temperature races down at sunset. Viking I measured a typical variation at Chryse, from −35°C at 6:00 P.M., to −50°C at 8:00 P.M., to −75°C at midnight. If such variations occurred on Earth, the average temperature would not be 15°C but would instead be well below the freezing point. Glaciers would sweep down from the poles, and the planet would be covered with ice almost to the equator. Why does this not happen? The reason is our much-maligned greenhouse effect. During the day sunlight heats the Earth's surface. At night, the warm ground releases this heat as infrared radiation, but it does not escape from the atmosphere as it does on Mars. The Earth's atmosphere contains several greenhouse gases that will not let heat escape into space at night or at least will considerably impede its escape. The most important of these gases are water vapor, carbon dioxide, and ozone, all of which absorb infrared radiation efficiently. As the most abundant, though, water vapor is the most important. Witness the fact that in desert regions, where the air is dry, the temperature drops sharply at night, sometimes to below freezing: the Egyptians made ice by leaving trays of water outside at night to freeze.

Without this natural greenhouse effect, which traps the heat in at night, the average temperature of the Earth's surface would be about −20°C. Although we hear much alarming talk about the greenhouse effect, without it human beings could not inhabit this planet. In fact, any life larger than bacteria would have extreme difficulty surviving, let alone flourishing. Given that when the Sun was newly created, it was perhaps 25 percent less luminous than it is now, it is possible that without the benefit of the greenhouse effect, life would never have started at all.

A Poor Greenhouse

Not only does Mars have a thin atmosphere, but, in contrast to Earth, there is only a little water vapor in that atmosphere. Even though two important greenhouse gases exist on Mars—the water vapor and also carbon dioxide—they do so in quantities too small to trap much heat. At present, the Martian atmosphere raises the average temperature of the surface by only 5°C. It is a pretty poor greenhouse. But why "at present"? Because this situation can change. The inclination, or obliquity, of Mars's axis of rotation is currently 25°, which is similar to that of Earth, and its orbital eccentricity is 0.093. But the orbit of Mars can be greatly influenced by the presence of Jupiter. At present the eccentricity is close to its maximum value, although it will decrease as the orbit becomes more circular. At the same time, the inclination of the planet's axis of rotation can range from 12° to 38°, which is a pretty impressive tilt.

These changes have a huge effect. What happens when Mars tips so far over? The poles then receive much more heat from the Sun during the summer. In summer the pole cap will melt more completely, and water vapor and carbon dioxide will be released in much larger quantities than at present. The atmospheric pressure will rise, and the degree of greenhouse effect will increase considerably. This in turn will have two results. One is that the rise in temperature will liberate even more gas from the summer pole. The other is that because the winter pole will not be as cold, less of the liberated gas will freeze out of the atmosphere.

According to some estimates, the enhanced greenhouse effect could cause the atmospheric pressure to rise by a factor of six to eight—and if a significant amount of water is trapped in the planet's surface away from the poles, the effect could be even greater. Should large amounts of permafrost melt, this would release even more water vapor, thereby further escalating the greenhouse effect. Combine all these factors with an eccentric orbit and things could become interesting. A strongly tilted pole that points toward the Sun at perihelion would be especially good at producing a global greenhouse effect that by trapping and storing more heat would sharply reduce what we would otherwise expect to be much colder winters. But what about the opposite situation? Combine a pole that points almost straight up with a circular orbit and you have a worst-case scenario. Neither pole will ever warm significantly in summer, and the atmosphere will progressively freeze out. As this happens, even the minimal greenhouse effect that currently exists will diminish, reinforcing the overall process in much the same way as above. Some estimates suggest that in such a situation as much as 99.99 percent of the already thin atmosphere could freeze. It may be, then, that Mars has a cyclic climate, with warm and cold phases each lasting a few tens of thousands of years. Until the present cold spell ends and Mars leaves its current ice age, Goldilocks will find the planet far too chilly for her taste.

A Greenhouse Gone Mad

In contrast, Venus has too much of a good thing. The planet is in the grips of a runaway greenhouse effect, which should serve as a terrible warning of what could happen to our Earth. When the solar system formed Venus was probably a warm planet, not unlike the Whipple-Menzel waterworld described above. Current theories about the formation of the solar system suggest that Venus may have started out with 50 percent more water than the Earth—it probably had oceans even more extensive than Earth's—and nine times as much as Mars. In fact, Venus could have had enough water to produce an atmosphere five times thicker than its current atmosphere, if all of the water vapor had vaporized at once.

Where is this water? Over the first thousand million years of Venusian history a "wet" greenhouse effect was in operation, and so the temperature of the planet rose. As a result, water evaporated from the oceans, passing into the atmosphere. The increase in water vapor enhanced this natural greenhouse effect, and so the temperature rose still further. As the air temperature steadily climbed, the temperature of the oceans continued to rise, causing more and more water vapor to enter the atmosphere, making the greenhouse effect ever stronger. Eventually the oceans evaporated altogether, and the surface of the planet turned into a steam bath. From there on the outcome was inevitable. The temperature rose inexorably until the carbon dioxide that on Earth is trapped in rock (as limestone and other carbonate rocks) was itself released into the atmosphere. The carbon dioxide reinforced the greenhouse effect still further, and finally a runaway point was reached. The water was lost to space as the surface of the planet heated to an extreme degree, until in the end all the carbon dioxide had been baked out of the rock, leaving the inferno we know now.

In other words, Venus is too hot because it has too much greenhouse effect, and it has too much greenhouse effect because positive feedback forced the process out of control. According to some estimates, this hellish scenario might have been repeated on Earth if it had been just 3 percent closer to the Sun.

Could the Earth Become Venus?

Could the same thing happen to the Earth? Prophets of doom claim that it is already happening, in the form of global warming and related changes in climate. The Earth's climate, though, is highly complex, and we do not yet understand everything about it as well as we would like. There is nonetheless good reason to believe that the Sun and solar activity affect the Earth's climate. We know that the Little Ice Age in the seventeenth and eighteenth centuries coincided with a period when the Sun was almost entirely lacking in sunspots and, surprisingly, was actually slightly fainter than it is today. Whether coincidentally or not, what with the Earth warmer now than it has been for several centuries, solar activity is at the highest point ever recorded. We also know that in the remote past the Earth's atmosphere was richer in carbon dioxide than it is now, without disastrous results ensuing—at least as far as we know.

What happens if the amount of carbon dioxide in the atmosphere increases? Carbon dioxide is a greenhouse gas, and so the temperature will rise with it. How-

Triumph or Disaster?

ever, carbon dioxide is a natural product of both volcanic activity and animal life, and the Earth has a system for scrubbing the atmosphere and removing the carbon dioxide. Part of it dissolves in the oceans, part is absorbed by plants, and part goes to form more carbonate rocks. The problem is, though, that if the Earth's temperature begins to rise, the oceans will get warmer and will be unable to absorb as much carbon dioxide. Some experts argue that plant life will take up the slack—that because more carbon dioxide will be available for plants to absorb, the vegetation will be lusher and therefore an even more effective scrubber of carbon dioxide. This process would reverse the increase in the amount of carbon dioxide in the atmosphere and the corresponding rise in the planet's temperatures. Others, however, are of the opinion that this compensatory process will not take place, or at least not to a sufficient extent, and that even if the effects were not as serious as on Venus, the rise in temperature could bring about cataclysmic changes in climate.

The whole issue is tremendously complex, and we do not understand all the checks and balances as well as we should. Some scientists hold that an increased level of atmospheric carbon dioxide is actually a good thing for the Earth—that it may be the only thing standing between us and a new ice age. Then again, some scientists warn us that unless we take immediate steps to *reduce* the amount of carbon dioxide in our air, we are courting catastrophe. The main question is, of course, whether the natural equilibrium of the Earth's processes will be able to cope with the growing impact of human life on the planet. Or, as Hollywood's *The Day After Tomorrow* suggests, will our lack of concern for the consequences of our behavior cause the entire system to spiral out of control? It is quite possible that we will not know the answer to this question for at least a century. A hundred years from now, though, it could be too late to remedy matters if the answer is not to our liking. Massive and rapid changes in the amount of greenhouse gases in the Earth's atmosphere may seem to some like a good idea, but we really do not know what the result of such changes might be—and it could turn out that the balance of factors that made our Earth "just right" for life is more delicate than we would prefer. My own feeling is that tinkering with things we do not properly understand is a recipe for disaster.

Popular Articles and Books

Patrick Moore, "Mars: Living or Sterile" (1969) and "Mariners to Mars" (1969), both in *The Sky at Night 3* (London: British Broadcasting Corporation, 1970).

> *Two short historical articles written in 1969 for Moore's popular BBC television series. Again, it is fascinating to see just how much things have changed, although some of the theories that Moore discusses have since been confirmed and are now solidly established.*

Patrick Moore, "The Hot Oceans of Venus" (1988), in *The Sky at Night 9* (London: Harrap Books, 1989).

> *An excellent essay on the wet greenhouse theory of Venus's evolution. Short and easy to follow, without any sacrifice of scientific accuracy.*

V. A. Firsoff, *The World of Mars* (London: Oliver & Boyd, 1969).

> *Val Axel Firsoff was an excellent speaker and lucid writer on Mars, who was willing to entertain ideas that were at or beyond the fringe of standard science. His book offers a detailed discussion of the greenhouse effect that is as good as any of the more recent descriptions and in fact better than most. Some of his speculations on climate also continue to be fascinating and quite plausible.*

More Advanced Reading

James B. Pollack, "The Atmospheres of the Terrestrial Planets," in J. Kelly Beatty and Andrew Chaikin, eds., *The New Solar System*, 3rd ed. (Cambridge, Mass.: Sky Publishing, 1990).

> *One of the definitive articles on the evolution of the atmospheres of Venus, Earth, and Mars. Pollack covers orbital variations, the greenhouse effect, and changes in the climate and surface conditions of these planets over the history of the solar system.*

On the Internet

The Cambridge Conference Network
http://abob.libs.uga.edu/bobk/cccmenu.html

> *Over the past two years CCNet has concentrated to an increasing degree on issues of climate research and climate change. Many of the articles are written from a skeptical point of view that readers may not necessarily agree with. All the same, they are aimed at making people think deeply about the complexity of such issues and the fact that much of the data is, to a greater or lesser degree, ambiguous.*

Going to the Planets?

Many possible disasters could overwhelm our planet, both natural—the impact of an asteroid, for example, or another ice age—or man-made, such as nuclear war, famine, a massive epidemic, pollution, or overpopulation. The possibility of such cataclysms brings to mind a comment made early in the twentieth century by Konstantin Tsiolkovskii, a largely self-taught Russian schoolmaster, who was the first to envisage modern liquid-fueled rockets: "Earth is the cradle of the mind—but you cannot live in the cradle forever." To put all our eggs in one basket—to limit ourselves to a single planet—is to risk that the human race may eventually be wiped out by some sort of catastrophe. Certain areas of the planet are already enormously overcrowded, which puts them at special risk. Were a large asteroid to fall in the center of Australia, several million people might be killed. Were one to fall in the middle of Europe, however, or in India or China, the carnage would be inconceivable.

Contrary to what many people think, exploring the planets is not merely a matter of scientific curiosity or purely theoretical research. Rather, it could ensure our own survival if the worst happens. Sending colonists to the planets would not only enable us to search out new resources but also to decentralize the human race such that a planetwide disaster would not spell the end of our species. The odds are utterly minute, but there are certain catastrophic events that would have an impact on the entire solar system, such as a sudden change in the Sun or a nearby supernova (although the nearest star that could turn supernova—Betelgeuse, in Orion—is actually far enough away that it would not do lethal damage). The more we spread out, though, the less likely it will be that a single disaster could annihilate us.

The colonization of the solar system is a standard theme in science fiction. Over the generations writers have carried their readers to the Moon, Venus, Mars, and the outer planets, and some have gone so far as to imagine the colonization of the Galaxy in a distant century. Some have populated the planets with a wide variety of unlikely inhabitants, whereas others have envisaged us swarming all over the solar system, setting up colonies and modifying planets to suit our taste. But just how realistic are these stories? Could we ever provide a hostile planet with a breathable atmosphere? Could the human race one day find itself living not only on the Earth but on other planets in our solar system—to say nothing of establishing a great galactic empire of the sort Isaac Asimov created in his *Foundation* trilogy?

Fiction or Fantasy?

I grew up reading science fiction. The first science fiction novel I read was *Mission to Mercury*, by Hugh Walters, which was among the titles on display when the local library mounted a presentation of children's books at my primary school. *Mission to Mercury* prompted me to read the rest of Walters's books, and other works, by writers such as Captain W. E. Johns, Patrick Moore, and Murray Leinster, followed.[1] Many of these books concerned the exploration and conquest of the planets. Sometimes the plotline and the science were believable; more often they were not. A few of them were so thoroughly absurd as to beggar belief. It is not a good sign when a ten-year-old critic can pick holes in the plot and the technical material!

Of course, I am far from the only science fiction aficionado. Ask the average person in the street whether they like science fiction, and chances are they'll say something like, "Yes! I just loved *Star Wars!*" No doubt, *Star Wars* was a wonderful film (as were its sequels), and one I thoroughly enjoyed—but it is not science fic-

y

Triumph or Disaster?

tion. *Star Wars* is a fantasy, or a space Western, set in a distant and exotic future. The film is not especially worried about whether this future is believable. To cite two of the most obvious examples, would Luke Skywalker really be able to shoot down ships traveling close to the speed of light with a hand-operated turbolaser? And in all those scenes in the hanger decks, how are people managing to breathe when the deck is open to space? In truth, much of what is billed as science fiction is fantasy or space opera—occasionally and somewhat barbarically called "science fantasy," even though it has little, if any, genuine scientific content. Rather, the hallmark of fantasy is its willingness to depict circumstances that defy the established laws of science and that we therefore know to be impossible. Much as I enjoyed *Star Trek*, the fact is that, despite its pretensions to being intelligent science fiction, the adventures of Captain Kirk and the green-blooded Mr. Spock were as much the stuff of pure imagination as the scenarios conjured up in the truly abysmal *Battlestar Galactica*.[2]

A true science fiction story is a "What if?" It takes an idea—for example, "What if we can develop a really efficient system for converting mass into energy?" or "What if we were able to travel faster than the speed of light?"—and explores its possibilities. It tries to work out what the consequences might be if these possibilities came true. Of course, when we read novels about the exploration of the solar system, we must remember that not all writers are concerned first and foremost about whether what they are writing is altogether scientifically plausible. Their interest may lie more with depicting a world that might result from our conquest of the planets and what might happen to their characters in such a setting. That said, good science fiction writers generally do try to base their stories on solid scientific fact, as opposed to idle speculation, whether they are working with elements of present-day science and technology or extrapolating from them to what *may* one day be possible (although they are not usually claiming that this is the way things *will* be). In what follows we will discover to what extent science fiction writers have been on the right or the wrong track about what may be one of the most critical issues the human race will face over the next few centuries.

Do We Belong in Space?

Is it worthwhile to send men into space at all? Why not send robot probes instead? These are important questions. At least for the purposes of initial exploration, it is cheaper, simpler, safer, and more cost-effective to use automated probes. In fact, in many such cases it would be too difficult, or dangerous, or expensive, or just plain ridiculous to send astronauts. But until we have developed far more sophisticated computers than exist today, programmed with systems for self-repair and endowed with the capacity for judgment and serious decision making, it is difficult to argue that humans have no place in the exploration of space. A simple system failure that an astronaut would be able to fix in a matter of seconds could wreck a mission, such as the pin that jammed, thereby preventing the Galileo probe's main antenna from deploying, or the "Abort" alarm light on the lunar module that switched off only when one of the Apollo XI astronauts gave the panel a sharp slap. An astronaut is also able to take action on the spot, whereas ground controllers must wait for data from an automated probe to be transmitted. The signals from a Mars probe, for example, may take as long as twenty minutes to reach Earth. Even from the Moon, the delay is two and a half seconds. This means that a robot moon rover must move very slowly to allow the operators sufficient time to respond. But imagine trying to pilot a Mars rover equipped with a half-hour delay on the steering! A robot rover could become mired in sand or fall over a small precipice before the controllers could see that a problem lay ahead.

Naturally, the more sophisticated the mission, the more a human presence is required. No robot probe could have carried out the exploration of the Moon that the Apollo astronauts did. Four Soviet robot landers were able to collect a few samples of moon rock, but they had to do so blind, and in the end they added almost nothing to the priceless knowledge gathered by the Apollo missions. Only someone with truly infinite faith in technology would imagine that the question of past or present life on Mars could ever be settled by robot landers. Similarly, to be effective, any large-scale geological or other scientific survey must be carried out by astronauts. Try programming a robot to decide which rock is the most interesting, or to figure out the solution to a problem, or to be curious or intuitive. Saying that there is no place for manned spaceflight is like telling Columbus that he was unnecessary—that his ships could have discovered America all by themselves.

A Home in the Sky?

Before we can colonize the planets, obviously we have to reach them. The human race made its first tentative steps into space in the 1960s and then withdrew. In recent years, however, interest has shown signs of reviving, and one hears rumors that the green light for a mission to send astronauts to Mars is imminent. The most optimistic speak of 2020. As part of its Aurora program for the manned exploration of space, the European Space Agency (ESA) talks of the possibility of putting someone on Mars by 2025. My own opinion is that 2030 is more realistic, and it may not even be much before 2040. One problem is that manned flight to Mars will be extremely expensive, at least if it is carried out with today's throwaway technology. Even so, assuming that we do not withdraw from space yet again, the person who will be the first to walk on Mars is probably alive today.

As we saw in chapter 5, in the late 1960s NASA had plans to build a manned moon base within the next decade and hoped to put men on Mars by 1980. By the early 1990s, however, NASA officials were forced to admit that if they were ordered to return astronauts to the Moon within ten years, they would be unable to do it, whatever efforts were made and whatever the budget available. Nor has the situation measurably improved. Some forty years of cost cutting and budget slashing, to say nothing of the burden of an ever-expanding bureaucracy that has filled NASA with administrators at the expense of scientists and engineers, have taken their toll. Today, we are unable to do in ten years what NASA managed in only eight years, and from a standing start, following Kennedy's declaration in 1961 that the nation should set itself the goal of landing a man on the Moon and bringing him safely back home. And yet the Moon is the key to the entire solar system. Without a return to the Moon, any further exploration of space becomes immensely more difficult and costly. In fact, if we are to reach the planets, returning to the Moon is essential, and for reasons that are not just a matter of psychological motivation. As we will see, establishing ourselves on the Moon is a necessary first stage in the technology we will need to go further afield. Any journey begins with a single step, and the step from the Earth to the Moon is vital to our venture into space.

Is a return to the Moon plausible in the medium term? If only out of self-interest, our answer has to be "yes." As resources run out on Earth, we will have only two options. One will be to search for ways to ravage our planet even more thoroughly than we have to date; the other will be to tap into the huge resources that the solar system has to offer. Within a few months' flight from Earth—that is, only slightly further away than the New World was for the Pilgrim Fathers—is more mineral wealth than we can imagine and plenty of space for setting up mining operations and industrial plants, which would then no longer need to deface the Earth. As some have said, out there in space it is raining soup, and we are so stupid that we haven't yet invented bowls. Soon we will have to decide whether to sacrifice what little of the Earth is left unspoilt or to look into space for what we need. The Moon may prove to be a short-term answer to many of our requirements for raw materials. Beyond the Moon, though, the asteroids offer metals and possibly even more valuable resources such as water, as well as a means to refuel in space, which would cut both the cost and the difficulty of interplanetary travel manyfold. However, to make this jump into space, we will first need a foothold on the Moon.

Why the Moon? The main answer is economics. The Moon will save us money. One of the greatest surprises since the space race was the realization that a lunar base *could* be self-sustaining. Even our science fiction stories have seldom, if ever, imagined such a development. The assumption has instead been that a lunar base would have to depend on the Earth for basic supplies such as air and water. Science fiction writers have typically envisaged such a base as an extension of the Earth and therefore as suffering from the same shortages and the same rationing programs as an increasingly hungry Earth. Instead, recent discoveries make it seem possible that a lunar base could not only be self-supporting but even highly profitable.

The Apollo missions proved that the lunar surface has a high proportion of valuable minerals. A typical Apollo XI rock sample is about 11 percent titanium oxide, 10 percent aluminum oxide, and 7 percent magnesium oxide, and titanium, magnesium, and aluminum are all very useful metals for engineering and construction. The discovery of large concentrations of titanium in lunar rocks—some samples contain as much as 13 percent of titanium oxide—was not something we anticipated. Another extremely valuable lunar resource is silicon. Lunar rocks contain 40 percent or more of silicon dioxide, and pure silicon is the raw material used in solar cells that convert sunlight into electricity. With two weeks of uninterrupted sunlight each month, the Moon is a paradise for the generation of solar-electric energy—energy that could be used on the Moon or beamed to Earth by tight-beam microwave transmitters. Lunar exports of solar energy could therefore massively reduce the Earth's need for fossil fuels and nuclear power.[3] Given almost limitless solar energy, it would also be a relatively simple job to set up automated factories to extract the silicon dioxide from the rock and mass-produce solar cells that would be able to generate any amount of power that might be needed. Such automated factories could convert large areas of the lunar surface into solar power farms just by using a small amount of the solar power already being generated to produce additional solar cells, in an exponential process. Moreover, once the silicon dioxide had been extracted from the rock, it could be broken down, thereby freeing up large quantities of oxygen—another necessary resource for a manned base, of course.

Prospecting for Ice

As we saw earlier, the Apollo program demonstrated that the Moon has almost everything necessary for human habitation other than hydrogen. The moon rocks were totally desiccated, and they showed not the slightest trace of any minerals containing hydrogen. Evidently, then, all the water for human or industrial use, as well as hydrogen for rocket fuel, would have to be supplied by the Earth, at crippling expense. However, recent observations have radically revised this conclusion by detecting what seem to be huge quantities of ice at the lunar poles. This discovery stems directly from one of the most unexpected findings in recent astronomical history. In 1993, radar observations of Mercury made from the Arecibo Observatory, in Puerto Rico, revealed that the polar regions of the planet appear to be extremely reflective. The simplest explanation for this would be that large amounts of ice are lying in shadow at the bottoms of craters: ice is highly reflective of radar waves. The fact that one of the hottest objects in the solar system could have ice caps ranks as one of the most remarkable developments in the modern history of the exploration of the planets. But there is a simple explanation. This ice is probably the result of the collision of thousands of comets with the planet. Every time a comet hit the surface, the ice and frozen gases in the comet would form a temporary atmosphere, part of which would freeze at the poles. Following the impact itself, dust and bits of rock would slowly settle, covering the ice and forming a mix of ice and rock, which would protect the ice from evaporating. The ice that remained in shadow could survive for millions of years, and thus, over time, large deposits of ice could build up. If this happened on Mercury, might it not also happen on the Moon? If such ice did exist on the Moon, it could easily be recovered by using open-cast mining to collect the ice-bearing rock and then by using vacuum stills to separate the water from the rock.

The first indication that the Moon might have ice at its poles came from a probe called Clementine. Clementine was a lunar mission by accident. It was funded by the Pentagon, and its main role was as a test bed for military technology. But as a lunar probe it did a fine job. Launched in February 1994, it had, among other tasks, the mission of mapping the south pole of the Moon. Of particular interest was a region close to the pole called Luna Incognita, which cannot be readily observed from Earth and had never been satisfactorily mapped by earlier space probes. What scientists had in mind was the possibility that, especially in view of the many high mountains in the region of the south pole, there might be areas of permanent shadow where ice still survived. This ice would have been produced in the same way as the ice on Mercury—by comets smashing into the lunar surface over the ages.

Clementine did indeed pinpoint areas at the bottom of craters around the south pole that never see sunlight. In addition, the probe discovered regions around the south pole that are highly reflective to radar. These coincided with regions in permanent shadow and were thus strongly suggestive of the presence of ice, although not proof of it. All the same, the information provided by Clementine produced much excitement. Perhaps these areas would turn out to hold the one resource that was missing from the Moon. Then, in December 1996, the Pentagon announced that ice had been found. The evidence was sufficiently unclear, however, that many people remained skeptical. All the same, the news was highly instrumental in getting the Lunar Prospector mission funded.

The Lunar Prospector mission was launched in 1998, and, as its name implies, its key aim was to locate resources, including ice, that would be of use to a future moon base. In the end, the probe was deliberately crashed into the Moon's south pole, the hope being that the impact would send up a cloud of water vapor, thereby proving that ice did indeed exist at the pole. This hope was disappointed, but before it met its end the Lunar Prospector probe had mapped the surface gravity and the Moon's magnetic field and had also produced detailed maps of the distribution of mineral resources such as iron, thorium, and potassium on the lunar surface. These elements can be identified because they emit X rays of a characteristic energy, which serve as a fingerprint. Not all elements emit X rays that can be detected, however, especially not the lighter elements, and hydrogen—the one element of serious interest that failed to appear in any of the samples of moon rock recovered by the Apollo missions—is the lightest of the elements. But hydrogen can be detected in another way. When solar and cosmic radiation bombard the Moon, they cause neutrons to be ejected from the lunar surface. Hydrogen absorbs neutrons, which is why it is used, in the form of water, to stabilize nuclear reactors that are getting out of control. Lunar Prospector was able to measure the number of neutrons emitted from the lunar surface as the result of its bombardment by radiation from space. Because cosmic radiation comes equally from all directions in space, the number of neutrons emitted by the lunar surface should be roughly the same everywhere. A drop in the number of neutrons suggests that something, possibly hydrogen, is absorbing them. Therefore, if the regions in shadow near the Moon's south pole proved to emit substantially fewer neutrons, this would be strong evidence of the presence of hydrogen in the surface rocks, presumably in the form of ice.

On December 2, 1996, NASA scientists announced the results of the Lunar Prospector probe. As had been hoped, there was a substantial region around the south pole, coinciding with the regions of permanent shadow, that emitted 25 percent fewer neutrons than did the rest of the Moon's surface. But scientists had not anticipated that areas of permanent shadow, where ice possibly survived, might exist at the north pole as well as the south, and so it was to the surprise of everyone when the probe identified an even larger region around the north pole that was deficient in neutrons. Specifically, Lunar Prospector found that there were two spots close to the lunar north pole where the number of neutrons emitted dropped by about 10 percent more even than at the south pole. The areas where the absorption of neutrons is the most intense are not located precisely at the poles but rather correspond to the areas in permanent shadow. Although some people have expressed doubts, if the increased rate of absorption is indeed due to water, then it appears that the Moon has a sufficient amount of ice to supply a fairly large lunar colony with water for several centuries. According to the lowest estimate, there are 10 million tonnes of ice mixed with fragments of rock and moon dust. If we make a conservative assumption that it would cost $100,000 to convey one kilogram of water to the Moon using the technology available today—that's $100,000,000 per tonne—the value of this ice to a lunar colony would be an astonishing $1,000,000,000,000,000,000 (one thousand trillion dollars).

What's more, careful analysis of the data now suggests that the initial estimates of the amount of ice were too low and that, rather than 10 million tonnes, there could be as much as 6 billion tonnes of ice at the lunar poles. Although much of it is mixed in with the lunar soil (the regolith), with perhaps as little as 1 percent ice in the mix, according to this new analysis there must also be regions of almost pure ice. To put this figure in proportion, 6 billion tonnes of ice is equivalent to an ice cube measuring 1.8 kilometers along each side. Such quantities of water would be enough to supply even a large human settlement on the Moon for thousands of years.[4] If the Lunar Prospector figures are correct, then even given a highly wasteful consumption of 100 liters of water per person per day, there is enough ice in the lunar poles to support a colony of a thousand people for 160 million years. Assuming that the discovery of huge quantities of ice at the lunar poles is confirmed, it is thus reasonable to believe that a self-sustaining lunar base is possible, one requiring only minimal resupply from the Earth. The water trapped at the poles could be used for growing crops and could also be broken down into hydrogen and oxygen, to provide hydrogen fuel for rockets and, as with the splitting of silicon dioxide, oxygen to breathe.

One Small Step to the Moon

In 1968, Arthur C. Clarke presented a paper to the Fourth International Symposium of Bioastronautics and the Exploration of Space, held at Brooks Air Force Base in San Antonio, Texas. He pointed out that, in strict energy terms, carrying an astronaut to the Moon is cheap: it requires the equivalent of about a thousand kilowatt-hours of energy to lift an astronaut—no rocket, just the astronaut—to the Moon. That is roughly the amount of electricity that a 100-watt lightbulb burns in a year, which, at 1968 prices, cost around ten dollars. Even allowing for inflation, the cost in 2004 is not very much higher. Why, then, is each launch so much more expensive? At present, a ticket on the space shuttle is priced at several million dollars, and the cost of launching payload is estimated to be an exorbitant $11,000 to $14,000 per kilogram.

Getting to the Moon is expensive not because of the distance but because every gram that goes to the Moon must be lifted against the Earth's gravity. The Moon is, on average, 380,000 kilometers away, but the costly part of the journey is the first 200 to 300 kilometers. In the case of Apollo XI, for example, lifting a total of 54 tonnes into lunar orbit required a rocket full of fuel, which meant that a heavy load of fuel was needed to lift that rocket into space, on top of which even more fuel was necessary to lift the heavy load of fuel. In the end, the Saturn

Triumph or Disaster?

V rocket weighed more than 3,000 tonnes, 95 percent of which was needed for the first few hundred kilometers, until Earth orbit was achieved.[5] A mission to Mars, however, will require a spaceship capable of transporting supplies sufficient to maintain a crew of perhaps half a dozen people for a minimum of two years. The ship will also have to bring along a landing vehicle, which, given that the astronauts will no doubt plan to remain on the Martian surface for at least several weeks, will need to be stocked with plentiful supplies for the landing party. It will also need to be designed to operate under the gravitational conditions on Mars. Because Martian gravity is much stronger than the Moon's, the landing vehicle will need to be considerably larger than the lunar module and will necessarily contain a greater amount of fuel—and so, in turn, a heavier initial load of fuel will be required to get both the spaceship and the landing vehicle into orbit around the Earth. It all adds up to a far bigger spaceship than the tiny Apollo moonships.

If we were to begin our journey to Mars from the surface of the Earth, we would therefore be looking at launching a minimum of several hundred tonnes, and quite possibly several thousand tonnes, into Earth orbit, something that would require a whole fleet of rockets as big as the Saturn V. Thus, although launching a manned Mars flight from Earth is perfectly feasible, the economics are appalling. In contrast to the Earth, however, the Moon has a weak gravitational pull: it takes just a twentieth as much energy to launch a rocket from the Moon as from the Earth. If a mission to Mars could leave from the Moon's surface, even a small spacecraft could carry enough fuel for the trip. Moreover, it would be possible to reserve a large fraction of the payload for cargo: only a small part need be fuel. That is why the lunar module of the Apollo landings could be such a small vehicle. Little weight of fuel was needed to carry two men and their supplies down from lunar orbit for three days on the Moon's surface and even less to lift them back into orbit in just the upper stage of the lunar module. In short, the key to reaching the planets is the Moon. If we can build rockets on the Moon and launch them from there, costs will be vastly reduced, which makes inexpensive interplanetary travel a real possibility.

To use the Moon as a base for exploring the solar system, however, we also have to be able to get to the Moon efficiently. With giant throwaway rockets launched directly to the Moon from the Earth's surface, the Apollo landings were anything but cost-effective. If we hope to make practical, long-term use of space, we will have to change the way that things are done. The space shuttle is a first step on this road. As we have seen, the hardest part of a journey to the Moon is getting men and supplies into orbit around the Earth. Doing so requires big rocket ships. It cannot be avoided. However, if we need only to get these spaceships into orbit, rather than all the way to the Moon, we will save a lot of weight and thus money. Large cargoes can be carried into low Earth orbit by rockets of a size that would struggle to send just a few kilograms all the way to the Moon's surface. If we are serious about going to the Moon, our initial destination would therefore be a space station in Earth orbit. Large rockets would be necessary only for going back and forth between the Earth's surface and the space station. Passengers would then transfer to other vehicles for the ride to the Moon itself.

Getting from the space station to the Moon would require nothing more than a small shuttle rocket. Moreover, if we had a second space station in lunar orbit, we could make the moon shuttle simpler still. The trip from Earth orbit to lunar orbit could be made in a pure space vehicle that would cruise on a trajectory requiring little fuel. Since it would never descend to the Moon's surface, it wouldn't even need a landing gear, which would save additional weight. From lunar orbit a third ship could be used that would simply blast between orbit around the Moon and the lunar surface. Each ship could be specially designed for one segment of the journey, which would maximize efficiency. In this way, moon flight would become truly cheap and easy—and this, in turn, would open the way to the planets. No interplanetary probe would ever need be launched from the Earth again. Instead, all manned planetary spaceflight would start from the moon base.

Skeptics will suggest that spaceflight is so ridiculously expensive that these ideas are pure fantasy: there never will be manned travel to the planets because it is just not worth it. This ignores a basic truth, one that is almost always forgotten in the debate about the space program. Space travel is expensive because we are still buying the technology. Once the technology is in place, the basic cost of a manned flight to Mars will be relatively trivial. As I pointed out earlier, this investment in technology is not, as some ill-informed critics suggest, a matter of money thrown down the drain. Who could have imagined that one of the most far-reaching benefits of the Apollo program would be that thirty years later hundreds of millions of homes and businesses would have personal computers and Internet access? What's more, these computers are many orders of magnitude more powerful than any of the computers of the 1970s, which filled a whole room. Whether directly or indirectly, the investment in electronics research alone has paid for itself many times over and has injected untold billions of dollars into the world economy.

Onward and Outward

It is no secret that the past several decades have witnessed a technological explosion, and yet as far as spaceflight is concerned we are primitives. Every flight demands the contribution of hundreds, if not thousands, of technicians and engineers, and even a trivial failure could mean catastrophe. Imagine if commercial aviation worked the same way, with pilots requiring months of training for each trip and the attention of a huge ground crew focused on just a single flight, with disaster lurking around every corner and the airplane thrown away afterward. Clearly, air travel would never have become practical.

In 1903, Simon Newcomb, a highly reputable Canadian scientist, claimed to have proved that heavier-than-air flight was impossible. On hearing of the Wright brothers' pioneering flight just a few months later, he was forced to concede the point—but, he argued, an airplane would never be able to carry a pilot and passenger. And yet commercial aviation was just twenty years away. Today, of course, aircraft routinely carry hundreds of passengers in relative comfort over thousands of kilometers. What is more, millions can afford air travel—a development that nobody imagined even at the start of the 1950s, when air travel was still limited to the wealthy. Just as this book was about to go to press, Sir Richard Branson, founder of the Virgin Group of companies, announced that he had contracted with Mojave Aerospace Ventures to produce five spacecraft that will be used to carry tourists into space, with the first such flight scheduled for 2007. For a week of preflight training and then a three-hour luxury suborbital trip that will include three minutes of weightlessness, prospective customers will pay roughly $180,000. But Sir Richard is confident that prices will eventually come down to the point where "masses of people" will be able to enjoy the experience of traveling into space.[6]

It is therefore not ludicrous to think that in fifty years space travel will be in the same state as air travel was in the 1950s—accessible to an ever-growing number of people, although still not quite an everyday occurrence. Burt Rutan, one of the two cofounders of Mojave Aerospace Ventures, is even more optimistic and envisages affordable space travel coming about in ten to fifteen years. But achieving this presupposes at least two important technological advances. As it stands, because fuel will not burn without oxygen, which does not exist in the vacuum of space, a rocket like the space shuttle or the Saturn V or the giant Russian Energia must carry huge tanks of liquid oxygen in addition to its own fuel. But this is a ridiculously inefficient way to proceed, given that for much of the journey into space the rocket is surrounded by air—that is, by oxygen. Ordinary aircraft engines are able to make use of the oxygen in the atmosphere up to an altitude of at least 20 kilometers (the altitude at which the Concorde used to fly). If rocket engines were adapted so that they could do likewise, traveling up to an altitude of some 30 to 40 kilometers by drawing oxygen from the atmosphere around them before gradually switching over to their internal oxygen tanks, a huge saving in weight and therefore cost would be made.

The second critical technological advance will be to design orbital shuttles capable of taking off from and landing on standard runways, with little more preparation and technical support than a typical intercontinental passenger flight today. Such a design may sound like fantasy, but it is not. Incredibly, for more than twenty years a design has existed for a winged air-breathing rocket that would be able to use ordinary runways: the HOTOL design. HOTOL (Horizontal Take-Off and Landing) would completely revolutionize spaceflight, and most experts believe that it is the way to proceed. There is only one problem: the design is British. The project was undertaken jointly by Rolls-Royce and British Aerospace in 1982, but in 1989 the British government, scarred by the Concorde project, canceled its funding. The design was highly regarded by the ESA, but Britain's lukewarm participation in the ESA is a major handicap to British projects being taken up for development. The cancellation of the HOTOL project has probably set back any serious commercial exploitation of space by perhaps as much as thirty years.

Once in space, where oxygen from the air is no longer there for the taking, essentially the same chemical rockets that have been in use since World War II would be adequate for the remainder of the flight into orbit and then onto the Moon. (The HOTOL spacecraft, for example, would use liquid hydrogen as fuel.) For interplanetary flight to be feasible, however, we will need to do much better. One of the chief arguments against manned flight to the planets has to do with the time required for the journey, rather than its intrinsic difficulty. For one thing, the longer the flight, the more mass that must be carried in supplies. For another, months of zero gravity are seriously prejudicial to the human body, causing severe calcium loss in the bones as well as the deterioration of muscle tissue. There is also the danger of radiation exposure. If astronauts were to travel on the traditional eight- to nine-month slow transit to Mars currently used by unmanned probes, they would have reached the maximum permitted lifetime dose of radiation by the time they returned to the Earth. Added to this is the ever-present danger of a solar flare, which could kill the astronauts in a matter of hours. It is a little-known fact that a massive and possibly lethal solar flare occurred just a few days after one of the Apollo missions returned to Earth. Had the astronauts been caught on the Moon by an unexpected flare their only chance of survival would have been to blast off into lunar orbit at once and turn the service module toward the Sun in the hope that its bulk would provide sufficient protection from the radiation.

But in fact there is no need for such lengthy transit times. Engines powered by atomic energy offer the possibility of continuous thrust and therefore much faster journeys. An atomic engine uses a nuclear reactor to heat a liquid "fuel" (technically a misnomer because it is not burned) to such a high temperature that it vaporizes and then ejects the gas at enormous velocity. The greater the velocity of the exhaust gases, the greater the velocity that the space vehicle can attain. The best fuel would be hydrogen, as it is the lightest element of all, but liquid hydrogen is bulky and difficult to manage. Helium is worse because it only liquefies at temperatures close to absolute zero, which makes it awkward to handle and to store in a spacecraft. Both ammonia (NH_3) and methane (CH_4), however, contain large amounts of chemically bound hydrogen in an easy-to-use form, which would make them a sensible alternative for deep space missions. At the temperatures—perhaps around 6,000°C—that should be attainable in an atomic engine, methane gas would in any case immediately break down into its constituent atoms of carbon and hydrogen. We could even consider using another simple compound, water, but it has less hydrogen and is thus less efficient.

Although an atomic engine would not be able to generate more than an extremely low thrust—perhaps only a tenth of a gravity, or roughly a fiftieth of what the Saturn V was capable of—it could sustain this thrust for weeks. How fast would our spaceship be going after three weeks of accelerating at one-tenth of a gravity? The answer is astonishing: 1,800 kilometers per second, or well over 150 times the escape velocity of the Earth. In that time the spacecraft would have traveled 1,700 million kilometers, which would put it in the orbit of Saturn. Besides greatly reducing the medical problems associated with zero gravity, such a ship could reach Mars in less than a month, even allowing for deceleration halfway.

Atomic engines—sometimes called ion drive—are, in a way, another story of missed opportunity. NASA had a working design, the NERVA (Nuclear Engine for Rocket Vehicle Application), which used a nuclear reactor to heat liquid hydrogen. There were plans to use this engine either in the third stage of an expanded Saturn V rocket for a manned mission to Mars or in the Nova booster that was initially the rival of, and later proposed as the successor to, the Saturn V. The Soviet Union also successfully tested nuclear designs. The trouble was that, as originally conceived, spacecraft equipped with these engines would have been flown in the atmosphere, and this raised fears about accidents. In open space, though, nuclear propulsion is without doubt the most sensible way forward. In a test on the ground, a later version of NERVA ran nonstop for fifty hours, thereby demonstrating that rapid, continuous-thrust flight to the planets is far from a fantasy.

Atomic engines are not the limit of the story, however. The 1950s and 1960s, particularly the early 1960s, were a tremendously fertile period for ideas concerning advanced spaceflight. In fact, many of these ideas have never been improved upon. One of these was the Orion project. The project was classified, and even today little is known about it, although, interestingly enough, it has been featured in several science fiction novels.[7] The Orion system proposed using large numbers of nuclear bombs to power an immense spacecraft. In order to fly, the rocket would drop bombs, and the recoil from their explosion would then push the rocket along. Even though nuclear fission, which is based on uranium or plutonium, is relatively inefficient in comparison to nuclear fusion, it is still immeasurably more energetic than a chemical fuel and thus theoretically capable of lifting not merely 3,000 tonnes of Saturn V rocket but millions of tonnes.

According to the original proposal, the Orion would have launched from the Earth's surface, but the 1963 Limited Test Ban Treaty made this illegal. Moreover, ecological concerns argue against such an idea. A revised proposal was made to build an Orion spacecraft in orbit, to be used as a Mars exploration vehicle. It could make the round trip to Mars in four months, as opposed to the current eighteen months, carrying a crew of eight along with plenty of supplies. If, in the future, such a spacecraft were to be launched from the lunar surface, where it could be constructed from local materials, it would be an ideal form of transportation for colonization missions to Mars. In fact, the Orion would make manned missions to planets as distant as Saturn a realistic possibility because the time required for the journey would be greatly reduced. Another major benefit of the Orion is that almost half of the entire mass of the vehicle could be payload. Space travel could take place in spacious surroundings, and a large load of cargo could be brought along as well.

Will We Ever Colonize the Planets?

There is, of course, a world of difference between flying to the planets and colonizing them. At present, any plan to establish colonies elsewhere in the solar system must remain a distant possibility. The notable exception is the Moon, where the potential for commercial gain might hasten the process: it is no coincidence that the Lunar Prospector mission was so named. The union of commercial, scientific, and national interests could conceivably lead to a permanently manned moon base in fifteen to twenty years. That said, unless something intervenes to improve NASA's present circumstances—a large increase in its operating budget, for example, or technological developments that substantially reduce the cost of spaceflight, or a new political or military agenda—such a schedule could prove optimistic.

Leaving aside the Moon, however, things are very much in the balance. Much is made of the suggestion that the planets, Mars especially, could serve to reduce population pressure on the Earth. But this is not a feasible idea. According to the GeoHive Web site, as of July 1, 2004, the estimated world population was 6,379,157,361 and was increasing by 1.14 percent per year—that is, by 199,239 people per day. Even assuming that technological advances eventually make space travel as simple and affordable as commercial aviation is today, it is impossible to imagine that two hundred thousand people could be shipped to other planets each day or that these planets could absorb them, year after year after year. In short, to argue that colonies on Mars could serve to relieve population pressure on Earth is absurd.

We are, moreover, unlikely to be able to exploit Mars commercially anytime in the near future, given the well-nigh ruinous cost of getting there using present-day technology, nor does the planet currently have any obvious military value. At the moment, then, and in some ways regrettably, the only serious reasons for going to Mars are political or scientific. The most important scientific reason is, of course, that manned exploration is the only way we are ever likely to settle the question of whether life exists, or once existed, on Mars. But there are limitations to what we as a society are willing to spend in order to satisfy scientific curiosity. Without a new space race that makes getting to Mars ahead of everyone else a matter of national pride, or some other, unforeseen development, Mars is likely to remain on the back burner, at least until the technology of spaceflight changes radically enough that interplanetary flight becomes straightforward and inexpensive.

Similarly, in the nearer term, manned flights to Venus or Mercury seem out of the question. There is no reason to send astronauts to such inhospitable, even hellish, places, where simple survival would be a major challenge, and in any case these planets can be adequately explored by robotic probes, at least for now. Likewise, neither does there seem to be much reason to go to Uranus, Neptune, or Pluto. The first two are, for all practical purposes, smaller and more distant versions of Jupiter and Saturn and so do not seem to present any compelling need for manned exploration. As for Pluto, what with its tiny surface frozen to an almost unbelievably low temperature, it is, in its own chilly way, as uninviting an object as Mercury or Venus. However, with the technology that now exists or could soon exist— notably atomic engines, which would dramatically reduce transit times—we could reach Jupiter and Saturn. In fact, odd as it may seem, apart from the Moon, the most promising sites for development all lie beyond Mars.

Perhaps most attractive in terms of sheer commercial interest are the asteroids. A vast quantity of mineral wealth, including an abundance of metals, lies within them that it will surely be in our interests to exploit. Furthermore, although asteroids do not contain crude oil, it is not inconceivable that the carbonaceous asteroids (those rich in carbon) could someday be mined for organic compounds that would, when refined, supply us with the equivalent of petrochemicals. The best way to carry out such mining operations would be to establish permanent bases on Mars or on some of the larger asteroids themselves. In addition, Saturn offers the attraction of the dense, hydrogen-rich atmosphere of its moon Titan, which would make it an ideal place for refueling. Interplanetary travelers preparing to return home could stop there to refill their fuel tanks. Especially because Titan would probably be the solar system's one and only gas station, it is potentially a very valuable piece of real estate. Some science fiction writers have even imagined a future in which, space travel now a commonplace event, Saturn—with its spectacular rings—has come to be as popular and exotic a holiday destination as, say, the islands of the Caribbean are today. Saturn also boasts billions of tonnes of water in the form of ice in its rings. The availability of such a mammoth amount of water might even one day open the possibility of seeding Mars and Venus with water.

In the long term the Jovian system may also prove to be prime real estate. It is possible, as the science fiction author Robert Heinlein has done, to imagine a future in which part of the thick surface layer of ice on Jupiter's moons Ganymede and Callisto could be electrolyzed, thereby furnishing these two planet-sized satellites with breathable atmospheres. Vehicles such as the Orion could carry large loads of the materials and equipment needed to make such planetary engineering possible. What is less certain, at least for the moment, is whether we could succeed in artificially creating a sufficiently intense greenhouse effect to make these satellites habitable, without which a project for colonizing Jupiter's moons would make very little sense.

Planetary engineering is indeed a fascinating possibility, if a long-term one. Schemes have been devised that would provide both Mars and Venus with atmospheres adequate to support future human settlers. In the case of Venus, the primary aim would be to reduce the quantity of greenhouse gases, something that could be accomplished by seeding the upper atmosphere with photosynthetic algae. The algae would absorb carbon dioxide, causing the surface temperature slowly to drop as the level of greenhouse gases in the atmosphere diminished and more heat therefore escaped into space. But then Venus is also totally desiccated. Once the temperature had dropped sufficiently, it would be necessary to bomb the planet with comets or with icy asteroids to provide it with water—or, as we saw, the huge amounts of ice in the rings of Saturn could be put to practical use in this regard. Even though it would be several centuries before such a project produced significant results, if mankind were up to the challenge, the payoff would be tremendous.

A similar argument can be made about Mars, where it would be necessary to add greenhouse gases to raise the surface temperature, rather than the other way around. Here, though, we face an uncertainty. A lot of carbon dioxide may be locked up in the pole caps. If so, warming the pole caps slightly would release this gas, while at the same time the melting ice would produce another good greenhouse gas, water vapor. A process of positive feedback could thus be set in motion, leading to a comparatively rapid warming of the planet as the atmosphere thickened. If there is *no* reservoir of carbon dioxide, however, then raising the atmospheric pressure and increasing the planet's very weak natural greenhouse effect would pose a considerably more complex technological puzzle. One of the major goals of future Mars exploration will be to discover just how much carbon dioxide and water actually is frozen in the soil and in the poles.

To provide Mars with a surface atmospheric pressure even a fifth of Earth's—as some science fiction stories propose—would mean releasing a staggering 10^{15} (one thousand trillion) tonnes of gas, at the very least. This is equivalent to melting a cube of ice nearly 100 kilometers across or landing a thousand comets the size of Comet Halley on the planet. While theoretically doable, creating such an atmosphere on Mars is not something likely to be achieved even in several centuries. Moreover, in order for human settlers to be able to breathe this atmosphere, it would have to be virtually pure oxygen. A sensible first step, then, would be to aim for an unbreathable atmosphere rich in carbon dioxide on which plants would thrive. Such an atmosphere would also generate a strong greenhouse effect that would warm the planet and release the water frozen in the soil. As the vegetation slowly absorbed the carbon dioxide, oxygen would be released, which would very gradually transform the atmosphere into a breathable one, just as happened on Earth hundreds of millions of years ago. At the same time, water vapor would take over from carbon dioxide as the principal greenhouse gas, allowing the planet's warmth to be retained.

The Choice Is Ours

In 2001 the popular magazine *Psychology Today* featured an interview with Buzz Aldrin. As the interviewer noted, Aldrin holds firmly to the opinion that a serious effort should be made to promote space tourism. Getting ordinary people into space, he argues, is likely not only to hasten our return to the Moon but also to arouse greater interest in a manned mission to Mars. Aldrin is the founder of a nonprofit organization called ShareSpace, which is dedicated to encouraging mass-market space travel, and is also the author of *The Return*, a science fiction novel in which space tourism is depicted as a flourishing industry. "If he had his way," the interviewer explained, "large numbers of tourists would soon be floating around in zero-gravity," an idea described as "a little far out."[8]

That a plan for developing space tourism—one proposed by the second man to walk on the Moon, no less—could be characterized in a such a way testifies to just how badly our expectations and our faith in our capabilities have eroded over

Triumph or Disaster?

the past several decades. The sad truth is that the notion of colonies in space now strikes most people as ridiculous—the stuff of fantasy (or of science fiction). And yet nobody laughed in 1977, when Gerald O'Neill published *The High Frontier: Human Colonies in Space,* a work that envisions huge cities circling in Earth's orbit. In fact, I remember a BBC current affairs program devoting a large slice of time one night to the book and the ideas it contains—and nobody laughed at that either. Just over thirty years ago few people doubted that our future lay in space. NASA was able to launch as many as four manned moon shots in a single year. Commercial spaceflight seemed to be only a few years away, and adventurous civilians who could afford to do so began buying tickets for future moon flights. Plans also existed for a manned flight to Mars by 1980. In short, it seemed all but certain that we were on the verge of a huge leap forward in our conquest of space.

But, as we have seen, since then progress has been limited at best, and in some ways the situation has grown worse. Just to take one example, in the mid-1960s NASA could launch a Gemini spacecraft piloted by two astronauts into high Earth orbit. Gemini X reached an altitude of 763 kilometers and, in September 1966, Gemini XI flew to a then-record height of 1,370 kilometers. The Gemini may have been relatively low-tech, but it worked and was in fact surprisingly versatile. In contrast, as of 2004 it is only with major effort that the United States (or Russia, or China) can launch astronauts even into low Earth orbit. Although in theory the space shuttle can achieve an altitude of 1,000 kilometers, with minimum crew and no cargo, as far as I can ascertain the record to date is 620 kilometers, set by the STS-31R mission on April 24, 1990, when the Hubble Space Telescope was launched. In October 1985, the Space Shuttle Atlantis flew to a height of 515 kilometers, but most missions do not exceed an altitude of 400 kilometers. The shuttles have almost as much power as the Saturn V rocket, but because they are heavier and carry more cargo than did the Titan II missile used for the Gemini flights, they are not as capable of reaching high orbits.

Ironically, the space shuttle, properly known as the Space Transportation System (STS), was originally designed in part for the purpose of launching military satellites into high polar orbits. But what with various compromises along the way and the need to adopt major safety precautions, especially in the wake of the Challenger disaster, the STS now has the capacity only to attain low Earth orbit. Although the initial idea was that the STS could be launched from Vandenberg Air

Force Base on polar missions, this has never happened—much to the disgust of the military, given that low equatorial orbits are useless for spy satellites.[9] In short, if we wanted at this point to send a manned spaceflight into high Earth orbit, whether on a scientific or a military mission, we would not be able to do so. Nor is NASA now capable of launching a manned spacecraft onto a translunar trajectory—that is, it cannot even reproduce the circumlunar mission of Apollo VIII (never mind actually putting an astronaut on the Moon). Consider also that Russians fly resupply missions to the International Space Station using automatic drones and two-person Soyuz spacecraft, which are small and light but were originally designed to get a single astronaut to the Moon and back, which the shuttles obviously cannot. Sending one of the space shuttles even as far as the ISS is like hiring a luxury coach when an ordinary taxicab would often do just as well, and at a fraction of the cost. NASA desperately needs a modern-day version of the Gemini or the equivalent of the Soyuz—something that would provide a simple, cheap, and flexible alternative for manned missions, to launch small payloads and just one or two astronauts. There would then be no need to use one of the shuttles for every mission.

At the start of the third millennium we are thus at a crossroads. One path leads to the Moon and beyond. The other route would, at best, take us no further than near-Earth space and could well lead to a complete withdrawal from space. The first road may be challenging and costly, but by choosing to pursue it the human race will broaden its frontiers and gain new knowledge. The second essentially leads nowhere, other than to stagnation—as if Columbus had decided that discovering the new world was too much like hard work. My nightmare is that our retreat from space could become total, a possibility explored in Allen Steele's disturbing novel *The Tranquillity Alternative.* Steele envisages a future in which a dispirited and rundown NASA gives up on manned spaceflight and sells its moon base and its remaining space assets to a private foreign company that operates commercially. Our fleet of space shuttles is aging, and the spacecraft have proved much less durable than we had hoped. If no decision is made either to upgrade them or to replace them with a new design, manned spaceflight could essentially come to an end by the end of the decade. And yet in the long term the decision not to take up the challenge of the planets would be disastrous for the human race.

Then again, a new incentive for a manned space program could develop. As

Triumph or Disaster?

we saw, our renewed exploration of space might initially be motivated by commercial interests, notably the desire to harvest lunar resources or to mine asteroids for their minerals. Or perhaps the scientific urge to settle the question of life on Mars will prove to be irresistible. Furthermore, according to reports coming out of China that are based on interviews with space program personnel, the Chinese may attempt to establish a moon base as early as 2010, an ambition that could trigger a new round of international rivalry in space. The Japanese are also showing an interest in manned spaceflight, with some Japanese corporations contemplating the possibility of having the commercial exploitation of the Moon underway within the next couple of decades. Even India is planning to launch a probe to the Moon by 2008. If Western governments find the idea that Chinese or Japanese could become the working language on the Moon unacceptable, we could find ourselves back in the situation of the 1960s, caught up in a new space race. While in the eyes of the general public such a development might not be welcome, the fact remains that competition has always been good for progress. Just as the original space race got us to the Moon, a new space race could easily be the spark that sets off an explosion in spaceflight technology, with long-term benefits for all.

Regardless of the incentive, if our space program is given the chance to expand, it cannot but revolutionize manned spaceflight, transforming it from something costly, cumbersome, and within reach of only the select few into a practical and affordable method of transportation. Technologies already within our grasp could effect a change as dramatic as that from the fabric-covered biplanes still prevalent at the start of World War II to the forty-passenger, jet-propelled De-Havilland Comet, which made its first commercial flight in May 1952, to modern, wide-body jetliners capable of holding upward of four hundred passengers. In short, investment in space promises countless benefits to the human race, many of which are yet even to be imagined. The solar system is there for the taking, and as a species we would be foolish not to make use of it, especially because it could be our salvation. Let us hope that in our desire for short-term economies we do not abandon our long-term vision of the possibilities awaiting us in space. To do so would be to kill the goose that *will* someday lay the golden eggs.

Science Fiction Books

Numerous science fiction stories and novels focus on space travel and the future colonization of the solar system. Some are highly unrealistic—the stuff of pure fantasy. Others are serious extrapolations from the present to "how it may be," although in some cases science has caught up with these stories, at least to some extent. Here, I mention just a few of the best. My two favorites were written more than fifty years ago and are still readable forty-five years into the space age, which goes to show that the best science fiction writing, if not immortal, can have a long half-life.

Robert Heinlein, *Farmer in the Sky* (New York: Ballantine Books, 1950).

> *Still my all-time favorite. The book describes the early colonization of Jupiter's satellite Ganymede. Remarkably, even after the Pioneer, Voyager, and Galileo missions to Jupiter, the story is still not particularly dated, in part because it deals with the technology of a still-distant future but also because, even if its ideas are sometimes highly speculative, in terms of their scientific background they are for the most part solidly grounded.*

Arthur C. Clarke, *The Sands of Mars* (London: Sidgewick and Jackson, 1951).

> *This book about the colonization of Mars is a classic of its type, despite being more than fifty years old and having been written six years before the dawn of the space age. We now know that Mars is a far more extreme environment than Clarke depicts, with no possibility of advanced vegetation, but his tale is still eminently readable.*

Kim Stanley Robinson, *Red Mars* (London: HarperCollins, 1992).

> *The first and by far the best book in a trilogy about the founding of colonies on Mars. It sets the first manned Mars flight in 2020 and follows the fortunes of the follow-up flight that lands in 2027. Although the time frame is somewhat optimistic, the story makes highly plausible use of technology that existed when it was written and likewise draws on the best available scientific information in describing the surface of Mars and the conditions that the explorers will encounter. Robinson's novel also deals in more detail than the others with the psychological problems that the first settlers may experience, as well as other implications of their isolation from the Earth.*

Popular Books

Arthur C. Clarke, *Report on Planet Three and Other Speculations* (London: Corgi Books, 1971).

A collection of essays by the master of science fiction on a wide range of themes. Many of the pieces were written in the 1950s, before the advent of spaceflight; the more recent date from the time of the Apollo moon landings. Several of the essays explore the issues of manned spaceflight and the possible colonization of the planets in considerable detail, but they do so in a thoroughly readable fashion, sometimes in the form of short stories. Clarke manages to explain many fundamental truths about spaceflight in a way that the average reader will have no difficulty following.

Gerald K. O'Neill, *The High Frontier: Human Colonies in Space* (New York: William Morrow, 1977; repr. Garden City, N.Y.: Anchor Books/Doubleday, 1982).

In this case the colonies take the form of immense habitats in Earth orbit that generate solar power for our planet, but much of the technology is based around the aggressive exploitation of the Moon. Superbly researched and beautifully written, O'Neill's volume holds out one vision of how the colonization of space could begin. The fact that so long after O'Neill's death, in 1992, the book is still being reprinted with the addition of material by such outstanding physicists as Freeman Dyson is testimony to its importance.

Patrick Moore, *On the Moon* (London: Cassell & Co., 2001).

A wonderful book that is in fact an extensively reworked version of Moore's Guide to the Moon, *originally published in the late 1940s and revised on numerous occasions up to 1977. Written by one of the last great astronomers to work from visual observations of the Moon, this is a complete, and easily accessible, guide to the Moon that covers all the modern discoveries and also includes a great deal of historical information. The final chapter focuses on the possibility of lunar colonies. Although Sir Patrick is very much a skeptic when it comes to some of the current theories about the Moon—for example, he only recently accepted that the lunar craters were caused by the impact of asteroids—his account remains authoritative.*

On the Internet

Encyclopedia Astronautica
www.friends-partners.ru/partners/mwade/spaceflt.htm

An exhaustive Internet encyclopedia of spaceflight that includes details of boosters, engines, and specific projects, and all the many variants on a particular design devised for different missions.

HOTOL

www.friends-partners.ru/partners/mwade/lvs/hotol.htm

> *A brief history and description of HOTOL offered by the Encyclopedia Astronautica site. Includes a low-resolution image and some basic specifications.*

James Garry's Interim HOTOL Page

www.fastlight.demon.co.uk/jg/hotol.html

> *A useful page with some artistic renderings of HOTOL by an aerospace expert.*

The Orion Project

www.astronautix.com/articles/probirth.htm

> *An article that describes the history of the Orion nuclear-bomb drive from its initial conception through its various design stages. A source of many fascinating details about the project and about how the Orion could serve as a vehicle for interplanetary travel.*

CHAPTER 12

Are We Stardust?

P oets and romantics often say that we, the human race, are stardust. What few people realize is how literally true this is. A few years ago I appeared in the BBC documentary series *Earth Story*.[1] During two days of remarkably hard work at the Roque de los Muchachos Observatory in La Palma (that amounted to just fifty seconds of film!), the question arose of whether we truly are stardust. Although at first glance a human being may appear to have little in common with a motorcar or a supertanker, this is not quite true. Both the car and the tanker are largely made of iron, just as the hemoglobin that makes our blood red and allows us to absorb oxygen from the atmosphere contains iron—and that iron was created in the center of the most massive stars in the Galaxy. In fact, most of the material in our bodies and in the planet we live on is made from stuff that was produced in the interior of stars much larger and more luminous than our Sun. We are children of the stars in a far more fundamental sense than most of us can imagine.

In the Beginning

You, me, this book, the chair that you are probably sitting in, and everything else in the universe is built up from a small number of different kinds of atoms called elements. Every schoolchild learns about the periodic table of the elements and how they are arranged in groups that have similar properties. It is commonly stated that there are ninety-two naturally occurring elements, number 92 being uranium.[2] The first and lightest of the elements, hydrogen, is extremely common: it makes up about 75 percent of the atoms in the universe. Helium, the second lightest and second most common element, accounts for another 22 percent, so that hydrogen and helium together make up 97 percent of the material in the universe. All the other, heavier elements—carbon, nitrogen, oxygen, magnesium, aluminum, sulfur, potassium, calcium, iron, copper, zinc, silver, tin, iodine, tungsten, gold, mercury, lead, uranium, plutonium, and so on—are contained in the remaining 3 percent.

It might seem logical to suppose that all the elements were formed at the same time and have always existed in more or less the same proportions as they do today, but this is not the case. Immediately after the Big Bang happened, there were no elements at all. First, the subatomic particles—neutrons, protons, and electrons—condensed from the primeval soup. Then, as the cosmos cooled, these particles formed into atoms. The simplest atom is hydrogen, in which a proton and an electron meet and are attracted by their opposite charges, so that the electron starts to orbit the proton. Most of the protons and electrons in the universe joined up in this way to form hydrogen atoms. But with a huge number of particles crowded together in a universe less than three minutes old that had barely begun to expand after the initial explosion, it was also easy for two protons, two neutrons, and two electrons to come together to form helium atoms. Helium is also created naturally in the classical nuclear fusion reaction such as takes place in the Sun, in which four hydrogen nuclei—that is, four protons—combine to form one helium atom, which has two protons and two neutrons, liberating energy in the process.

You might assume that to produce the elements that follow helium in the periodic table—lithium (with a nucleus containing 3 protons), beryllium (4 protons), boron (5 protons), and so on—all you need to do is keep adding protons. Not so. Instead, when a further proton is added to helium, the result is so unstable that the series never has a chance to continue building. In the Big Bang, the construction of elements ceased as the density of the explosion dropped—and at that point, when the elements stopped forming, just three minutes into the history of the universe, what had been produced was a lot of hydrogen, about a third as much helium, and a small amount of deuterium, or heavy hydrogen. There were, as far as we know, no other elements, which leaves us with a rather considerable puzzle. As is plain to see, even a small planet like the Earth contains untold billions of tonnes of heavier elements. What's more, on Earth helium is actually quite rare: it exists as a tiny fraction of our atmosphere, released by radioactive elements that are trapped underground in gas pockets.

If we look at our own bodies, we can appreciate the problem. About 80 percent of the human body consists of water. Water is, of course, made up of two atoms of hydrogen and one atom of oxygen, which weighs sixteen times as much as each hydrogen atom. Our bones are principally calcium and phosphorous. Our blood is based around iron. The proteins that we are made of contain hydrogen, but they are also composed of carbon, nitrogen, and oxygen. Wherever we look we find elements that could not have formed in the Big Bang. So where did these elements come from?

Was the Earth Born from the Sun?

The composition of the Earth is obviously unlike that of the Sun. The most common elements on Earth are silicon, iron, and oxygen. In contrast, about 93 percent of the atoms in the Sun are hydrogen, and the other 7 percent mostly helium, along with trace amounts of heavier elements. The most common of these heavier elements are, in order of their occurrence, oxygen, carbon, neon, and nitrogen—although oxygen atoms are only one-hundredth as abundant as helium, and nitrogen atoms ten times rarer still. At the same time, helium atoms have a mass four times that of hydrogen. By mass, then, the Sun is only 77 percent hydrogen but 21 percent helium, with the trace elements—which, despite existing only in very tiny amounts, are considerably heavier than either hydrogen or helium—now making up the remaining 2 percent. As we will discover, the Earth differs from the Sun in composition principally because when the solar system formed, the Earth, being small and close to the gravitational pull of the Sun, could not retain the hydrogen-rich gas that originally surrounded it. However, this begs an important question. Where did the Earth come from in the first place?

To answer this question, let us look at what goes on inside our Sun. What the Sun basically does is convert hydrogen into helium. There are two main ways to produce helium: the proton-proton (p-p) chain and the carbon-nitrogen-oxygen (C-N-O) chain. The simpler of the two processes—and the one that takes place in our Sun—is the p-p chain.

The nucleus of a hydrogen atom consists of a single proton, orbiting which is an electron. Inside the Sun, the immensely high temperatures shear the electrons off the hydrogen atoms, leaving them as loose protons, each of which has a positive electrical charge. Although under ordinary circumstances like charges repel each other, given the enormous temperature and pressure at the center of the Sun, two protons can be forced together. When this happens, one of the protons is converted into a neutron, which has no electrical charge, and the neutron and proton then combine to form deuterium. This we write as:

$$p + p \rightarrow d + e^+ + \nu$$

In addition to the deuterium atom (d), a positively charged particle called a positron, also known as an antielectron (e^+), is formed that carries off the electrical charge of the proton that has become a neutron, and with the positron a ghostly, almost undetectable neutrino (ν), to balance nature's books.[3] We are now able to detect these neutrinos from the Earth, and it is by analyzing their characteristics that we know that the Sun follows the p-p chain of nuclear reactions, not the alternative C-N-O chain.

Deuterium (heavy hydrogen) reacts rapidly. A deuterium atom, the nucleus of which contains a neutron as well as a proton, will collide with a proton to form helium-3, or light helium (^{3}He), which has two protons but only one neutron in its nucleus:

$$d + p \rightarrow {}^3He + \gamma$$

Energy is liberated in the form of a gamma ray. Helium-3 is stable, but it is a simple matter for two atoms of helium-3 (each with its two protons and one neutron) to collide to form the nucleus of an ordinary helium atom (two protons and two neutrons) plus two loose protons:

$$^{3}\text{He} + {}^{3}\text{He} \rightarrow {}^{4}\text{He} + p + p$$

The overall result is a fusion reaction in which four protons combine into the nucleus of a helium atom. The helium atom is slightly less massive than the sum of the four protons, though, and the lost mass is converted into energy according to Einstein's famous equation $E = mc^2$. This reaction consequently liberates a huge quantity of energy.

Up to now, the Sun has been "burning" (that is, fusing) hydrogen. When roughly 7 percent of the Sun's hydrogen has turned into helium, the core of the Sun will start to choke with what is effectively helium "ash," just as the ash from a coal fire builds up until it finally suffocates the remaining coal. In other words, the helium atoms begin to get in the way of the hydrogen atoms, thereby impeding the fusion reaction. Think of a quarterback who is being harassed: if he is surrounded by too many defense players, he will probably not be able to complete his pass. Likewise, with all that helium around them, the protons are unlikely to connect successfully with the deuterium nuclei. When this happens, a crisis will occur. As the fusion of hydrogen gradually abates, the nucleus of the Sun stops generating enough heat and energy to sustain itself against the force of gravity, which is attempting to crush it under the weight of the Sun's outer layers. With the Sun's nucleus weakened, the contest has only one possible victor. Suddenly, the nucleus of the Sun will collapse like a house with overstressed foundations. This collapse ratchets up the pressure on the nucleus considerably, and, as when any gas is compressed, causes the temperature of the nucleus to rise enormously.

What happens then? As we have seen, hydrogen reacts quite easily to form helium. But helium is far more reluctant to react and will do so only at phenomenally high temperatures and under a huge amount of pressure—the very conditions that the collapse of the Sun's nucleus creates. At the temperature at which the fusion of hydrogen was taking place—around a mere 15,000,000°C—the helium in the Sun was totally inert, but once the nucleus collapses and the temperature rises sufficiently, the helium can begin to undergo fusion. Helium is followed in the periodic table by lithium, beryllium, and boron, and so one might assume that the fusion of helium atoms will produce these elements. But lithium, beryllium, and boron are all extremely unstable. At temperatures well below that of the Sun's nucleus, they are immediately broken down again by stray protons, which crash into them like cannon shells. In other words, as far as creating new elements is concerned, combining two helium atoms achieves nothing, for the result immediately comes apart again into two helium atoms. To achieve something, we must jump over lithium and beryllium and crush together *three* helium nuclei.

Fortunately, then, there is this alternative path. Three helium nuclei, otherwise known as alpha particles, can join together to form a carbon nucleus, which has six protons and six neutrons:

$$3\ ^4\text{He} \rightarrow\ ^{12}\text{C}$$

The fusion of helium thus creates an additional element. But, unfortunately, even though helium refuses to react very readily, when it does react, it "burns" much more quickly than hydrogen and therefore gives the Sun only a temporary reprieve. When the helium in the Sun's core runs out, the nucleus will collapse yet again. This time there is no escape. The Sun is not massive enough that the collapse of its nucleus will generate a temperature high enough to permit two carbon nuclei to combine and thereby allow yet more elements to be produced. At this point, with nothing left to "burn," the Sun will collapse into a carbon-rich white dwarf star.

Thus, even though we have succeeded in producing some new elements, other than a little nitrogen and possibly some oxygen (both formed when protons crash into the carbon nuclei), we have nothing heavier than carbon. Moreover, the elements that have been formed are of little use because they are locked inside the white dwarf. We cannot get them out of the star and into the Earth. To do that something far more violent is needed.

Although our Sun is not massive enough to "burn" carbon, larger stars can—stars whose mass is, at minimum, around five times that of the Sun. Stars that size will happily combine carbon with hydrogen and helium, and with carbon itself, to form an assortment of new elements, the most common among which are nitrogen, oxygen, neon, and silicon. As each newer, heavier element is produced, it forms a layer at the center of the star. As we move out from the center, we find layers of progressively lighter elements in which fusion reactions are still occurring. Each time a heavier element is created, however, it sustains the star for a shorter time before the accumulation of unreactive "ash" strangles the process. The process of crisis, collapse, and renewed reactions thus continues, each new collapse bringing stability to the star for a shorter time than before as the temperature at the center of the star rises ever higher, allowing yet heavier elements to combine. Ultimately, in only the most massive stars, silicon forms, which heralds the end of the process. At temperatures over 300 million degrees, two silicon nuclei can fuse to form iron—but iron is the most stable atomic nucleus that exists. No reaction that involves iron produces energy: they all absorb energy.

At this point, then, we have an onionlike star consisting of layers in which different reactions are taking place. Right at the very center of the star, silicon is reacting to form iron. A little further out, where the temperature and pressure are not as high, large quantities of the lighter elements remain, which are continuing to undergo fusion. In one layer, carbon is still being converted into nitrogen, oxygen, neon, and so forth, and beyond that layer, helium is still combining to produce carbon, and in the outermost layer hydrogen continues to react to form helium. Beyond this layer, though, the temperature and pressure fall too low for fusion reactions of any kind to occur, and thus the fiery nucleus—which accounts for only about 10 percent of the star's diameter—is surrounded by a huge and for the most part tenuous atmosphere of hydrogen. But how does this solve our problem? Locked within the center of the star is a vast supply of elements such as carbon, oxygen, magnesium, silicon, sulfur, and iron, as well as many other elements less heavy than iron. The total mass of these elements may equal, or even exceed, that of our Sun. How, though, do we get these elements out of the stars and into space, where they can form planets like the Earth? The answer is surprisingly simple. Nature has found a way of recycling stars and spreading star-stuff out into interstellar space.

The Celestial Recycling Plant

What happens when a massive star has no material left to "burn"? On February 23, 1987, astronomers had the chance to see the answer. A young Canadian astronomer named Ian Shelton was working at the Cerro Tololo Inter-American Observatory at Las Campanas, in Chile. Using the observatory's 25-centimeter astrographic telescope, he took a photograph of the Large Magellanic Cloud, one of the two small galaxies that lie alongside the Milky Way. When he developed the photograph, he noticed a rather bright star that he did not recall having seen before. The object was immediately suspected of being a supernova. An excited panic ensued as astronomers scrambled to observe it, and within a few hours observers who had not yet heard the news were independently reporting the explosion. Obviously, something special had been discovered.

A supernova is the end of the life of a large star. There are two types. Shelton's supernova, the first of 1987 and thus named Sn1987a, was of the second type. A Type II supernova is at first sight less spectacular than a Type I, which emits more visible light. But a Type II actually releases much more energy overall, for it represents the death of a massive star of the multilayered sort just described, with a core full of iron. When the quantity of iron reaches a critical level, the nuclear reactions in the center of the star come to a halt, in the same way that a furnace choked with ash will eventually go out. The result is like taking away the star's foundations. Up to now, the fusion reactions have been generating the heat (that is, the energy) needed to counterbalance the tremendous force of gravity exerted by such a giant star. Once the reactions cease, there is nothing to prevent gravity from taking over. Just as an elevator that has suddenly had its cable cut plummets to the basement, the star collapses—but this time it has nothing to fall back on because iron will not take part in new reactions. In as little as a second, the core of the star implodes, and all the unconsumed elements in the outer layers suddenly start to combine in an orgy of nuclear reactions of all kinds. The outer layers crash into the center at velocities of thousands of kilometers per second, causing a rebound generated in part by the colossal increase in heat produced by the wave of uncontrolled nuclear reactions. A catastrophic explosion ensues, in which roughly 90 percent of the mass of the star is expelled into space at high velocity, leaving little more than the core of iron behind.[4] Measurements of Sn1987a show that the explosion expanded at around 17,000 kilometers per second, or some 5 percent of the speed of light.

As the supernova expands, it seeds the space between the stars with enormous quantities of heavy elements from the interior of the star. Many of the elements most common on Earth, notably carbon, oxygen, magnesium, aluminum, silicon, sulfur, and iron, are hurled into space. The explosion also creates elements that are heavier than iron, although these are not produced by means of nuclear fusion. Rather, the heavier elements result from the violent bombardment of existing elements, such as iron, by radiation. In the first moments of the supernova, when the expanding cloud is at an extremely high temperature, the frenzy of nuclear reactions taking place generates countless numbers of stray neutrons, in addition to all the protons and alpha particles (that is, the hydrogen and the helium nuclei, respectively). This incredible hail of particles bombards the nuclei of atoms of the existing elements, and these nuclei absorb some of the particles, thereby rapidly building much heavier elements, until the bombardment eventually ceases as the cloud expands and cools.

Many of the elements formed in this expanding envelope are extremely unstable and would, in the normal course of affairs, break down again very quickly. Such is the intensity of the bombardment, however, as well as the incredible speed at which neutrons, protons, and alpha particles are being absorbed, that the new elements do not have time to decay, no matter how unstable they are. In fact, there is good evidence that elements even heavier than uranium—the transuranic elements—may be formed in this blizzard of radiation. As might be expected, such elements are also highly unstable and will therefore decay radioactively, sometimes gradually and sometimes fairly swiftly, into stable ones. Unstable atoms can decay in one of two ways, either by giving off an alpha particle—that is, a helium nucleus, consisting of two protons and two neutrons—or by converting a proton into a neutron by emitting a positron, a process known as beta decay. In the first case, the atom attempts to become more stable by making itself lighter, while in the second it tries to achieve stability by giving itself fewer protons and more neutrons. This twofold process results in the formation of the so-called heavy elements (those heavier than iron) such as lead, tin, mercury, gold, and silver, which are found in the rocks of our planet, or the noble gases such as radon, a gas released by rocks like granite that contain substantial quantities of radioactive elements, or krypton and xenon, gases that occur in tiny quantities in our atmosphere. A good example of the process is the slow decay of uranium into lead, which requires that the uranium be transformed into a long series of intermediate elements before finally finding stability as lead. In the course of this process of decay, each uranium atom sheds a total of ten protons and twenty neutrons.

In short, a supernova is a factory for manufacturing elements and for turning one element into another. But it is also a recycling plant. It takes the elements that the dying star has used up, reprocesses them, and then spreads these elements into space, in much the same way that a household compost heap turns organic rubbish into valuable garden fertilizer. Once they have been essentially thrown out into space, these elements can someday combine to create, not carrots and squashes, but new stars and their planets. And so we see that all the many elements found on Earth originally formed in the interior of the most massive stars. How, though, does this material end up in a planet such as the Earth and in your body and mine?

When Supernovas Collide

To convert a stellar death into a stellar birth requires a stellar coincidence. All around the Galaxy, stars and planets form when clouds of dust and gas contract. But why should a cloud of this type start to collapse in the first place? The answer is that it may get hit by the force of a cosmic explosion, in the form of a supernova, which starts the process.

The Veil Nebula, located in Cygnus, appears to us as a luminous swirl in the sky, a large, slowly expanding circle of nebulosity. This is in fact a wave of gas and a little fine dust that resulted from a supernova some fifty thousand years ago. This material is still glowing dimly, mainly because it continues to collide with the interstellar medium—the thin gas and dust that fills the space between the stars. The force of the supernova's expansion scoops up and concentrates the interstellar gas and dust in front of the expanding wave of material from the supernova. When we look through a telescope, we can see the energy of the impact in the different colors of the nebula. At the outer edges, where the collision occurs with the greatest force, the level of energy is highest and thus appears blue, while the inner, more protected part, which is less energetic, has the characteristic red hue of gently glowing hydrogen atoms, from the hydrogen present in the nebula and in the interstellar medium. Of course, after fifty thousand years, this wave of gas and dust is now many light-years across. In a wide-angle view, we can see that, although some parts of it are much brighter than others, the nebula is in fact an almost complete ring, the pattern one would expect from an explosion that took place so very long ago.

Sometimes, however, a supernova explodes not in empty space but instead close to, or even inside, a gaseous nebula, that is, a dense cloud of gas that is ready to form stars. These dense clouds are stellar nurseries—accumulations of hydrogen gas left over from the formation of our Galaxy, from which new stars are constantly being born. As the expanding wave of material from the supernova explosion crashes into the gaseous nebula, it seeds the nebula with the elements necessary to form planets. Such is the case with the famous Orion Nebula, which can be just seen with the naked eye and is a lovely object when viewed through binoculars or a small telescope. In a nebula this large, many massive stars will be born. Initially, these stars tend to form in groups and, because they are massive stars, they have a short lifetime, sometimes only a few million years.[5] When these massive stars die, new supernovas occur. Each supernova not only further enriches the gaseous cloud of the nebula with heavy elements but also compresses it, sweeping the gas up in front of it just like the shock wave from an explosion, leading to the formation of additional new stars.

But how do we know that this is really what happens?

If our solar system was seeded by material from a supernova, then, in addition to the standard elements, there should also have been a large quantity of unstable, short-lived radioactive isotopes of these elements, which, as we have seen, are produced in the hail of radiation from the supernova. One in particular has attracted a great deal of attention—an isotope of aluminum known as aluminum-26, or ^{26}Al. Aluminum is usually composed of 13 protons and 14 neutrons and thus has an atomic mass of 27 (^{27}Al). The ^{26}Al isotope contains only 13 neutrons, however. Because it has too few neutrons for the number of protons it contains, ^{26}Al is unstable and highly radioactive—it has a lifetime of a mere 720,000 years—and decays into magnesium. In this process, one of the protons in the nucleus emits a positron, or antielectron, and, in so doing, loses its positive charge, thereby becoming an additional neutron. This leaves us with a nucleus containing 12 protons and 14 neutrons, otherwise known as magnesium-26, or ^{26}Mg. The most common isotope of magnesium is ^{24}Mg, which has 12 protons and 12 neutrons, but there are two other stable isotopes—^{25}Mg, with 13 neutrons, and ^{26}Mg, with 14 neutrons—although these are normally much rarer in nature. If we discover unexpectedly large amounts of ^{26}Mg, this can only be because a lot of ^{26}Al was once present but has since decayed, and the only conceivable source of large amounts of ^{26}Al would be a nearby supernova. This radioactive form of aluminum is an incredibly powerful source of heat. It is estimated that early in the development of the solar system, enough ^{26}Al existed to have melted small asteroids, up to perhaps 20 kilometers in diameter, by heating them from the inside, in much the same way that uranium and other elements heat the Earth's interior.

Similarly, a rare gas called xenon comes in many different types: there are no fewer than nine different stable—that is, nonradioactive—isotopes, of which ^{132}Xe is the most common. Tiny amounts of xenon are found trapped in meteorites, some of which is located in small white grains that meteoricists name "inclusions." These appear to include material that already existed somewhere in interstellar space before our solar system was created and was trapped when the meteorite formed.[6] When the xenon is analyzed—particularly the xenon found in the inclusions—something odd is discovered. Three of the isotopes are much more abundant than one would expect: ^{128}Xe, ^{130}Xe, and ^{132}Xe. What have they in common? These are all isotopes that, if we are correct, form deep inside red supergiant stars through the process of rapid addition of neutrons described earlier. Red supergiant stars are also the stars most likely to become supernovas. They are very massive—they often have twenty times the mass of our Sun, or more—and are large enough to "burn" elements all the way through to silicon, thereby winding up with the dead core of iron that ultimately triggers a supernova.

How to Make a Solar System

The Sun, the Earth, and all the planets formed from what we call the protosolar nebula, an enormous, dense cloud of gas and dust similar to the Orion Nebula. Although we say that the cloud was dense, were you inside it, you would think you were in a vacuum. On average, interstellar space contains about one atom of hydrogen per cubic centimeter, whereas something like the Orion Nebula typically contains thousands or tens of thousands of atoms per cubic centimeter. But even the densest nebulae have only a few million atoms per cubic centimeter—and a few million atoms per cubic centimeter corresponds to an extremely hard vacuum, better than anything that can be attained on Earth. By way of comparison, the lunar atmosphere has around two hundred thousand atoms per cubic centimeter, and yet it is so thin that it can barely be detected. However, when we take many cubic light-years of this apparent vacuum, the overall mass is huge—as massive as many stars the size of the Sun.

Somewhere around five thousand million years ago, the protosolar nebula started to collapse, probably, as we have seen, because a nearby star exploded as a supernova. Initially this process took place slowly, but it accelerated as the gas became denser. As the collapse of the cloud progressed, small internal movements of the gas picked up speed, setting the cloud spinning. The same effect can be observed in an ice-skater who begins a spin slowly, with arms extended, but, as the arms are pulled in, starts to spin more and more rapidly. In just the same way the cloud, which was once light-years in diameter, contracted to just a few hundredths of a light-year, and the tiny amount of movement it originally had gradually picked up speed. As the cloud spun more rapidly, it flattened out into a disk, growing hotter and denser at its center. Throughout the cloud small bits of material began to

Triumph or Disaster?

clump together, forming tiny bodies called planetesimals. These initially expanded by adhering to each other when they collided but later, as their gravitational pull increased, by attracting new material. In the densest regions of the disk these planetesimals grew rapidly. Over time, two in particular began to dominate and, through their superior pull of gravity, impeded the growth of others. These were the Sun and Jupiter. In fact, our solar system only narrowly avoided turning into a double star.

Initially dozens, or hundreds, or perhaps even thousands of planets formed. For hundreds of millions of years the solar system was a wild careening mass of bodies, some of them hundreds or thousands of kilometers in diameter, and many of them in highly eccentric orbits. Collisions between these protoplanets were frequent. It is almost certain that our Earth was destroyed by the impact of another planet, probably one around the size of Mars, following which the debris took shape again to produce the Earth-Moon system (as we saw in chapter 5). Another giant impact reduced Mercury to a small fraction of its original size, and some astronomers speculate that a similar impact may have split Pluto and formed its big satellite, Charon. On Mars, the Hellas basin, a giant crater some 2,100 kilometers across, was obviously created by the impact of an enormous object, while on our much smaller Moon, the same can be said for the large *maria,* or seas—the Mare Imbrium, for example, which is around 1,000 kilometers in diameter. Such impacts must have come close to breaking their targets apart. In other cases, relatively small objects collided with much larger planets, which were able to absorb the smaller bodies and thus increase in size. Jupiter and the Sun must have swallowed many of these protoplanets in something of a feeding frenzy. As a result, they became larger still, which in turn gave them an ever stronger pull of gravity with which to attract new material.

At the same time, a second effect was taking place. The dense rocky material and the heavier elements had been falling into the center of the revolving cloud. In the meantime, at a certain distance from the cloud's center, where the temperature had dropped considerably, the lightest materials, gases like ammonia (NH_3) and methane (CH_4), were freezing. In the hotter, more central part of the cloud, only the largest protoplanets, with their powerful gravitational pull, were capable of holding onto these hydrogen-rich gases, while in the outer regions were huge amounts of hydrogen-rich ices. Hence the protoplanets varied in composition. Closer to the center of the cloud were small, dense bodies, including the inner planets, made out of the rocky material that had not fallen into the Sun, and further away were bodies composed mostly of gas or ice. The largest of the distant bodies—Jupiter, Saturn, Uranus, and Neptune—had enough force of gravity that they could retain huge amounts of hydrogen gas, too, whereas the smaller ones consisted just of the ice.

Eventually, as the protosun got denser and hotter, the temperature at the center of the cloud increased to the point where a stable series of nuclear reactions could begin, and the Sun suddenly flashed into existence. When this happened, the Sun's radiation and the waves of hot gas escaping from its surface began to clear the gas and dust that was left over from the formation of the planets. Collisions between protoplanets swept up most of the larger material that still remained in unstable orbits and, by about four thousand million years ago, the solar system as we know it was essentially complete.

Given that the mass of Jupiter is about three times that of all the other planets put together, what was left has been described—very accurately, if unflatteringly—as the Sun and one large planet, plus some debris. We know it, somewhat less objectively, as the solar system, consisting of the Sun, eight major planets, Pluto, and untold millions (or perhaps even billions) of minor bodies that include planetary satellites, asteroids, and comets.

Children of the Stars

Are we stardust? The answer is most emphatically "yes." Look at the most common elements in our own bodies. These are hydrogen and oxygen, which make up the water that accounts for 80 percent of our weight, calcium and phosphorous in our bones, carbon and nitrogen in the proteins in our muscles, and iron in our blood. Of these elements, all except the hydrogen, which was formed in the Big Bang, were created inside stars. Iron forms only in the largest and most massive stars of all, as do phosphorous and calcium, which are intermediate steps in the building of iron. The lighter elements like carbon, nitrogen, and oxygen—the elements that are the basis of proteins—can form in less massive stars. But it is only in a supernova explosion, whether it be a tremendously violent Type II or the somewhat less energetic Type I, that any of these elements are liberated into space.

Iron is a life-giving element, for it allows our blood to transport the oxygen that we breathe. And yet it is also the poison that leads to the death of a massive star. It forms in the core of the star, but because iron cannot generate a new supply of energy, it causes the star to collapse and explode. In so doing, however, the star seeds space with all the elements that are required for life to begin. We thus owe our existence to the noble sacrifice of these great stars that, in their death, make our life possible.

SUGGESTIONS FOR FURTHER READING

Popular Books

Simon Lamb and David Sington, *Earth Story: The Shaping of Our World* (London: BBC Consumer Publishing, 2003).

> *Based on the 1998 BBC documentary series about the Earth and its history,* Earth Story *(also available under the title* Earth Story: The Forces That Have Shaped Our Planet*) traces the history of our planet from its formation up to the present day. Two centuries ago, scientists began to investigate the history of the Earth by examining the rocks beneath its surface, which led them to formulate the astonishing concept of geological time. Using this discovery as their starting point, the authors of* Earth Story *unravel the fascinating history of the Earth from its earliest beginnings to the dawn of human civilization. The writing is refreshingly straightforward, and the book is lavishly illustrated.*

Paul Murdin and Lesley Murdin, *Supernovae* (Cambridge: Cambridge University Press, 1985).

> *An excellent and very accessible discussion of supernovas—their importance and their consequences—written by a widely respected British astronomer, Paul Murdin, and his wife, who have collaborated on many books. The volume begins with a historical overview of the supernovas that have, over the centuries, been observed by the Chinese, Native Americans, Arabs, and Europeans. The authors then turn their attention to recent research and modern theories regarding supernovas.*

More Advanced Reading

Laurence Marschall, *The Supernova Story* (Princeton: Princeton University Press, 1994).

> *A more detailed account of supernovas, including what causes them and the effects they have on the Galaxy. Like the Murdins, Marschall examines the records of supernovas that have come down to us from our ancestors, as well as what we now know about these events. He goes on to present a full account of stellar evolution and the current theoretical models of what happens in a supernova, explaining how scientists test their theories against observations. Although not directed at a popular audience,* The Supernova Story *is not an unduly challenging read.*

Rudolf Kippenhahn, *One Hundred Billion Suns* (Princeton: Princeton University Press, 1993).

A comprehensive study of stellar evolution—how stars of various sizes develop, what happens to stars when they die, and why only some stars are likely to become supernovas. One need not be a mathematician to follow the material, but the nonspecialist reader will need to take a lot of the physics on faith.

On the Internet

Astronomy Picture of the Day
http://antwrp.gsfc.nasa.gov/apod/astropix.html

A site maintained by Goddard Space Flight Center that is a favorite of all astronomers, whether amateur or professional. Each day the site features a new photograph—almost always a spectacular one—that concerns some aspect of the universe. Many of the pictures have to do with the birth and death of stars, and all of them are accompanied by short explanatory commentaries. There is an extensive index, and one can also access thematic groupings of images. For example, images pertaining to the birth of stars can be found at http://antwrp.gsfc.nasa.gov/apod/stellar_nurseries.html.

Supernovae
http://imagine.gsfc.nasa.gov/docs/science/know_l1/supernovae.html

A first-rate overview of supernovas, with illustrations, that includes descriptions of the different types of supernovas and how they occur.

The JPL Infrared Astrophysics Team: MIRLIN Star/Planet Formation Page
http://cougar.jpl.nasa.gov/HR4796/anim.html

A page mounted by the Jet Propulsion Laboratory that, in addition to still images, offers an animation that shows how planets form, along with a clearly written step-by-step explanation of the processes involved.

Manned Spaceflight at the Crossroads

On January 16, 2003, the Space Shuttle Columbia STS-107 was launched. It was making its twenty-eighth flight. For the sixteen-day duration of the mission all appeared to be going perfectly. Then, just after 5:00 on Saturday afternoon, February 1, I connected to the Internet to check my mail, and there, in a message from a mailing list that focuses on spaceflight, I read the fateful words: "The Columbia has just broken up on reentry." For a moment I wondered whether this could be some joke, albeit in very poor taste, but both the BBC News Web site and, when I turned on the television, Sky News confirmed what I had just read. The shuttle had disintegrated, and all seven members of the crew were presumed dead. For more than seven hours time seemed to stand still as I sat riveted to the TV, only very briefly shifting my attention from the coverage of the disaster. Apparently the Columbia's heat shield had failed, although for reasons that were not immediately obvious.

In retrospect, alert observers might have taken note of two anomalous items of news—first, that during takeoff a small piece of insulating foam had fallen from the external fuel tank and then hit the shuttle somewhere on the left wing and, second, that radar had tracked a small object of unknown origin moving away from the spacecraft in orbit—probably a fragment of the damaged shuttle, we can now surmise. As it was, however, confusion initially reigned over the precise causes of the tragedy, although there was no shortage of speculation. Finally, seven months later, the Columbia Accident Investigation Board (CAIB) issued its report. The source of the tragedy was indeed the piece of foam insulation that had fallen off the external fuel tank during launch, evidently blasting a hole in the leading edge of the left wing. This allowed superheated gas to enter the wing during reentry into Earth's atmosphere, melting the wing from inside and destroying its aerodynamic stability. Ultimately, the wing broke off, and the spacecraft went into a violent spin before disintegrating and burning up in the atmosphere. Since CAIB's report was released, the three surviving space shuttles have been grounded until the recommendations of the board can be met, something that is proving to be a complicated and costly matter.

As of August 26, 2004, the first anniversary of CAIB's report, only five of the fifteen mandatory improvements—those that must be completed before the remaining shuttles can return to space—had been made. In particular, no way had been found of carrying out satisfactory emergency repairs in space to the shuttle's heat shield in the event of damage as severe as the 15- to 25-centimeter gash in the Columbia's wing—a problem that still remains unsolved. NASA's response to the danger posed by the foam has been to attempt to ensure that such an accident cannot happen again. The external fuel tank has been extensively redesigned in an effort to make it impossible for the insulation to fall off. Launches will now be tracked by additional cameras so that possible mishaps, such as the impact of debris, can be immediately detected. Further adding to this precaution, future launches will be carried out in daylight, at least for the time being. Astronauts will have the ability to repair minor cracks in the heat shield, and the hope is to have an inspection boom installed on the shuttle's robot arm so that the underside of the shuttle can be examined. But if major damage were discovered, the crew's only option would be to make an emergency rendezvous with the International Space Station and then await rescue. Therefore, a mission will be approved only if the spacecraft would be able, if need be, to make such a rendezvous. This has meant that the scheduled, and crucial, servicing mission to the Hubble Space Telescope, needed to ensure its survival, has had to be canceled because the orbit of the telescope is so different from that of the ISS that, were an emergency to occur, the shuttle would not have enough fuel to allow it to reach safety.

In the meantime, Russia's space program has in recent years been severely constrained by lack of funds and is reportedly almost bankrupt. At the time of the Columbia disaster, the Russians had not launched a cosmonaut for quite some time, and word was that they had only two Soyuz spacecraft left in reserve. Although they have since sent resupply and crew-change missions to the International Space Station, members of the ISS crew are now expected to remain in space for much longer than originally intended. The European Space Agency, which has also been beset with funding crises, suffered a major setback in 1992 when its design for the three-person Hermès space plane was abandoned after the cost of the project spiraled out of control, as did the weight of the vehicle. The increase in weight was caused partly by the addition of a number of extra safety features in the aftermath of the tragic explosion of the Space Shuttle Challenger in 1986, although some experts were already highly skeptical about the spacecraft's design and capabilities. The ESA's nascent program for manned space travel thus appears to have been suspended indefinitely, without a single mission ever having been flown. The sole exception to the trend is China. After years of intense speculation in the West about the status of the Chinese space program, in October 2003 the Chinese launched a manned spacecraft into orbit around the Earth. Since then, they have announced further plans for a moon landing within ten years.

In the United States, though, manned spaceflight has clearly reached a point of crisis. Critics argue that such programs are too expensive and too dangerous and that only unmanned missions should continue. The CAIB report sounds depressingly similar to the report issued in 1986, following the loss of the Challenger, and before that in 1967, after the tragic accident with Apollo I, and even, to a degree, after the explosion aboard Apollo XIII, which might so easily have culminated in catastrophe. The pattern is one of poor management, warnings that are ignored, and a tendency to hope for the best rather than take immediate action when trouble arises, often because that trouble is simply not recognized for what it is. When, for example, the ground crew at NASA reviewed the Columbia's launch and saw what had happened with the foam insulation, they concluded that the problem was unlikely to pose a grave danger. Above all, though, the evidence points to an appalling laxness regarding crew safety, an attitude that appears to be based in part on overconfidence and in part on poor communications within NASA. In the case of the Columbia, as the result of the breakdown in communications, warnings from engineers to the effect that the problem with the foam insulation was being underestimated did not get through to the senior managers who had the authority to take action.

Public perception—buoyed by the many successes of the space program over the years as well as by the image of a well-nigh infallible, "can do" NASA depicted in films such as *Space Cowboys* and *Armageddon*—has been significantly undermined by the news that NASA's initial assessment of the Columbia tragedy ("Yes, but even if we had realized the seriousness of the problem, there was nothing that could have been done") was incorrect. Had the potential risk not been dismissed, an emergency rescue mission might well have been launched sufficiently promptly. True, NASA would have needed to prepare the Space Shuttle Atlantis in record time, and, once notified of the problem, the Columbia astronauts would have had to ration supplies carefully for a longer-than-planned stay in space of perhaps as much as four weeks. Provided rationing had begun by day 7 of the mission, however, in all likelihood the Atlantis could have been underway in time—and the earlier a decision was taken, the greater the margin of safety would have been. The critics charge that once again NASA has demonstrated its incapacity to function efficiently and effectively and that over the years the agency has grown so accustomed to being mired in red tape that a fundamental change of mind-set is just not possible.

As British astronomer Heather Couper reminded us in an interview on Sky News shortly after the loss of the Columbia, NASA once declared that, should it ever lose one of its space shuttles on launch and another on reentry, it would abandon the design on the grounds that the vehicle was unfit for manned spaceflight. Now two of the original four shuttles are gone. The Challenger was lost on its tenth flight and the Columbia on its twenty-eighth, so the goal that NASA announced at the outset of the shuttle program—to make a hundred flights with each of its shuttles—has proved absurdly optimistic. Moreover, in 1986, when the Challenger was lost, funding was not nearly as tight as it is today, and replacement parts for the shuttle were still available, so that a new spacecraft could be constructed to take the Challenger's place. But no one seriously believed that the Columbia would be replaced. In fact, if CAIB's report had pinpointed a critical flaw in the design of the spacecraft, the three surviving shuttles would have had to be either redesigned, at considerable cost, or else scuttled, at which point it would have been all too easy to cancel the shuttle program entirely.

Whether under its present circumstances NASA's credibility could survive a repeat of the 32-month suspension of flights that followed the Challenger disaster is far from certain. We may shortly find out, though. As a target date for the next shuttle mission, NASA was originally aiming at the launch window that will open up in March–April 2005—already a good 26 months into the current moratorium. To meet such a date, NASA would have had to complete the remaining ten of CAIB's fifteen critical recommendations by December 2004. NASA administrators were bullish. In an interview with the Associated Press, for example, shuttle program manager Bill Parsons declared that NASA was steadily working toward a "return to flight" by the target date. All the same, there was no hiding the fact that progress had been slow. Then, at the start of October 2004, NASA announced that they would be unable to meet the March–April 2005 target date after all and were instead working to determine whether a launch during the window in mid-May would be feasible. As this book goes to press, NASA is aiming for a mid-May launch and is optimistic that this date can be achieved.

In January 2004 President George W. Bush made a major policy statement concerning the space program. He called for a return to the Moon, possibly as early as 2015 and no later than 2020. Research efforts would be aimed at allowing human beings to remain on the Moon for increasingly extended periods, partly so that the Moon's resources could subsequently be used in the service of explorations further afield, beginning with a manned mission to Mars. His proposal further stipulated that the United States would finish its share of work on the International Space Station by 2010, after which the facility would be used for long-term research into the effects of spaceflight on the human body. The existing space shuttles would be retired no later than 2010, and a new manned launcher, the Crew Exploration Vehicle, would be developed by 2008 and would serve in place of the present shuttle to transport scientists and astronauts to the space station. To put things in perspective, this would be the first new manned launcher developed by NASA since the early 1970s, when the current space shuttle was designed, and only the second in the past forty years (the other being the Saturn V rocket, designed in the early 1960s). At the same time, President Bush requested a substantial increase in NASA's budget, including an additional $1 billion over the next five years, and also proposed redirecting a further $11 billion of the existing budget into a program aimed at a return to the Moon. Grand visions are cheap, of course, especially in an election year, and many people tend to suspect that President Bush's proposals amount to little more than an electoral gimmick. But if so, then in view of the relative lack of public support for the space program, it is arguably a somewhat peculiar choice—hardly an obvious vote-winner.

In 1989, to mark the twentieth anniversary of the first Apollo moon landing, former president George H. W. Bush made a similar policy statement, setting forth a plan that would culminate in a manned mission to Mars. But Congress refused to fund his proposals, which were essentially thrown out the window. Fifteen years later, however, Congress is on the whole more supportive of the president. Even though the proposed funding increase for NASA has already been sharply curtailed, it has not been jettisoned entirely, and so it could be that President Bush's proposals will fare better than did those of his father, particularly if the American economy improves over the next few years. Moreover, the fact that the Chinese appear to be planning to put a man on the Moon provides a new incentive, for it seems very unlikely that the United States will settle for seeing another nation dominate space.

Sadly, of course, the loss of the Columbia was by no means the first disaster to take the lives of crew members. In January 1967 the three astronauts who would have flown the Apollo I mission were killed when a fire broke out in the pure-oxygen atmosphere of the command module during a preflight test. Only three months later, the Soviet Union lost a cosmonaut when Soyuz I crash-landed after having been plagued with problems from the outset of the mission, including a solar panel that failed to deploy and a series of serious navigational and instrumental difficulties. The pilot, cosmonaut Vladimir Komarov, consequently lost control of the craft and was forced to make a dangerous manual reentry, which he survived only to be killed after the main parachute refused to open and the backup one became tangled. Komarov's space capsule crashed into the ground at some 140 kilometers per hour and burst into flames. Three more cosmonauts perished in June 1971 when, owing to a faulty seal, the Soyuz 11 space capsule depressurized during its descent to Earth. The spacecraft itself landed intact, but when the recovery crew reached it, they found the cosmonauts dead in their seats. (Strangely, however, while the cause of death was officially given as asphyxiation, none of the three showed any signs of the struggle normally associated with such a death.) Then, as many of us vividly remember, in January 1986 the Space Shuttle

Challenger exploded shortly after launch, again killing a seven-member crew—although it is possible that most (perhaps all) of them survived the explosion. We know that at least one crew member was still alive and conscious inside the crew compartment as it plunged to Earth and smashed into the Atlantic Ocean and that several others had used emergency breathing equipment following the explosion. Dreadful though it is to contemplate, unless the Columbia went into an extremely violent spin after the left wing broke off—so violent as to knock the crew unconscious—it is likely that at least some of them survived long enough to realize what was happening to them. The fact that Mission Control received a brief burst of telemetry from the spacecraft some twenty seconds after voice contact was lost strongly suggests that the crew compartment remained intact.

In short, our quest for space has met with tragedy before. But we grow not by our successes so much as by our failures—how we overcome them and what we learn from them. As Rudyard Kipling famously counseled his son:

> If you can keep your head when all about you
> Are losing theirs and blaming it on you;
> If you can trust yourself when all men doubt you,
> But make allowances for their doubting too;
>
> . . . .
>
> If you can meet with triumph and disaster
> And treat those two impostors just the same;
>
> . . . .
>
> Yours is the Earth and everything that's in it,
> And—which is more—you'll be a Man my son!

No explorer, no nation, has ever become great by retreating in the face of difficulties. If we give up on space now, not only will we be turning our back on the future, but a great many lives will have been sacrificed in vain.

Notes

PROLOGUE

1. See the Bell Labs Web site on the discovery of the laser: www.bell-labs.com /about/history/laser/invention/invention9.html. Curiously, although both Townes and Schawlow figured on the patent, it was Townes who won the Nobel Prize in 1964, while Schawlow had to wait until 1981.

CHAPTER 1. Stonehenge

1. Despite being such an important figure in the history of Stonehenge, Sir Edward Antrobus appears to have left little mark on history. He was born in 1867, but apart from that I have been unable to locate much by way of information about him.

2. The plan is to completely redo the arrangements for public access. At present, a main road passes through the site, just a few meters from the stones. This road will be rerouted through a tunnel, while other roads will be removed and the land restored. A visitors' center will be constructed about 3 kilometers from the stones, and from there visitors will be able to gain access to the site only by walking or by taking "land trains" that will run on liquefied petroleum gas (a low-pollution fuel) and then walking a short distance to the stones. Even nearby plowed fields will be restored to grassland, so that Stonehenge will once again lie in an unbroken expanse of pasture and thus partly recover the appearance it must have had some four thousand years ago. The increased distance from the roadway to the site will also benefit security, given that people who have been out drinking have been known to pull over on their way home and climb the fence in order to carve a symbol of their eternal love for their chosen partner, or perhaps the name of their favorite sports team, on the ancient stones.

3. A tonne is a metric ton—which, as it happens, weighs almost exactly the same amount as a nonmetric ton.

4. To the best of my knowledge, the only person to connect Merlin with time travel has been T. H. White in the *Once and Future King*, who has Merlin living backward through time, growing younger as others grow older.

5. Some sources suggest that this eclipse occurred on October 22, 2136 B.C., but these suggestions appear to be based on a treatise on Chinese astronomy written in 1732 A.D. In fact, one can demonstrate that no eclipse took place on this date.

6. The red color is produced when the disk of the eclipsed Moon is lit by the light refracted through the Earth's atmosphere. In other words, the Moon's disk is, quite literally, lit by the red light of sunrises and sunsets all around the globe. Depending on how transparent the atmosphere is at the time, the color can be anything from a bright, coppery glow to a deep blood-red. The Moon rarely, if ever, disappears entirely during an eclipse.

CHAPTER 2. How Did the Stars Get Their Names?

1. The IAU created a definitive list of the constellations and their accepted names, established a formal definition of the boundaries between constellations, and stipulated that stars located on the border between two constellations must be assigned to one or the other of them. Today the IAU is responsible for the naming of all bodies in the solar system (planets and their satellites, asteroids, comets, and so on) as well as specific features found on them, such as craters, mountains, or volcanoes. The IAU has also created an officially sanctioned method for the naming of stars, galaxies, sources of X rays, and anything else in the cosmos, and is the final arbiter in disputes concerning names.

2. As formally defined today, a difference of five magnitudes—that is, between magnitude 1 and magnitude 6—is equivalent to a difference of exactly one hundred times in brightness. In most cases Ptolemy's estimated magnitude of each star is in close agreement with modern measures. A number of inconsistencies exist, however, some of which are so apparent that astronomers wonder whether the star has changed over the intervening centuries, since it seems unlikely that Ptolemy would have committed such an error. The star Sirius, in particular, has been the topic of much debate. Ptolemy described the star as red (*hypokirros*), whereas it is quite obviously white. In this instance, one suspects there genuinely was an error, possibly because Ptolemy's words were incorrectly transcribed. In two other cases, though, the stars may in fact have lost some of their brilliance since Ptolemy's time. Ptolemy rated the star Megrez, which is located in the Plough (or Big Dipper) as having the same brightness as the other stars in the constellation of Ursa Major, whereas today it is significantly fainter, well below magnitude 3. Similarly, Denebola, in Leo, which Ptolemy ranked as magnitude 1 and thus as one of the brightest stars in the whole sky, is now slightly below magnitude 2. Although we cannot be certain, it is possible that both these stars have faded over the past two millennia.

3. The use of Latin extended beyond the realms of scholarship. When George I, who did not speak English, ascended to the British throne, Latin became the language of government in

England. Cabinet meetings were, for example, conducted in Latin so that the king would be able to follow the proceedings.

4. We are referring here to the planets other than Pluto, which is a freak in that its orbit is much more strongly inclined than that of any other planet. Thus its path among the stars moves well beyond the ecliptic, and the planet can in fact appear in quite unexpected places. Although I regard Pluto with great affection, the news that it is passing through the constellation of Coma, for instance, largely fails to excite me. Presumably astrologers can also cope with this irksome aberration on the part of the smallest and most recent of the planets (if indeed it is a planet, for which see chapter 7).

5. The name 87GB073840.5+545138 can be broken down as the position of the source—073840.5+545138, that is, right ascension $07^h 38^m 40.5^s$ and declination $+ 54° 51' 38''$—in 1950 Besselian coordinates (B) in the 87GHz catalogue (87G) of the California Submillimeter Array in Hawai'i. This designation obeys the IAU's recommended conventions for naming to the letter.

CHAPTER 3. What Was the Christmas Star?

1. The *New Revised Standard Version* uses "at its rising" in place of the more familiar phrase "in the East" found in many other translations of the Bible. At Matthew 2:9, however, where the expression "at its rising" occurs again, the editors suggest "in the East" as an alternative.

2. Quoted by David Hughes in "The Star of Bethlehem," *Nature* 264 (1976), p. 513.

3. David Hughes argues that Luke's passage could indicate virtually any time of year except winter. On closer consideration, however, it seems far more likely that Luke refers to the spring than to any other season of the year. The shepherds probably would not spend the nights in the hills in summer simply because there would be no need for special vigilance at that time of year, and in autumn they would do so only if the sheep could not be brought down to shelter in a single day during the roundup before the start of winter. But the shepherds would need to be present at lambing time, in the spring.

4. The editors of the *New Revised Standard Version* in fact append a note to "wise men" in Matthew 2:1 ("wise men came from the East") suggesting "astrologers" as another possible translation. In most ancient and early modern civilizations rulers employed court astrologers whose job it was to interpret the signs observed in the sky and offer advice accordingly. This practice persisted in Europe up to the start of the seventeenth century, and in some countries it survives today. For example, in India and Sri Lanka the custom remains for elections to be called on the date that the government astrologer identifies as the most propitious.

5. Since this was written, astronomers have found a new asteroid, named 2004 MN4, which will pass just 3,000 kilometers from the center of the Earth on April 13, 2029. It will reach magnitude 3, thus being easily visible to the naked eye. Such an occurrence is exceptionally rare.

6. The three triple conjunctions in Pisces occurred in 980–79 B.C., in 861–60 B.C., and in 7 B.C. Other triple conjunctions took place in Leo in 821–20 B.C., in Taurus in 563–62 B.C., in Virgo in 523–22 B.C., and in Cancer in 146–45 B.C. The most recent were in 1940 and 1980–81, but there won't be another until 2238–39, followed fairly shortly by one in 2279.

CHAPTER 4. How Do We Know When Comet Halley Was Seen?

1. Many people are uncertain about the difference between comets and meteors. A comet is basically a large ball of ice. This is the nucleus of the comet, which can be as much as several kilometers in diameter. As the comet approaches the Sun, the nucleus heats up, and, as it does, part of the ice turns into gas. The gas then lifts up dust from the surface of the nucleus, and together the gas and dust form a tenuous, dusty atmosphere around the nucleus. If the comet gets close enough to the Sun, this atmosphere will be blown away from the comet by the solar wind, in the same way that wind blows smoke away from a chimney, thereby creating the comet's tail. In contrast, a meteor, or shooting star, is a tiny body, usually no larger than a grain of sand, that can be seen flashing across the sky as it falls toward the Earth and burns up in the atmosphere. Most meteors are grains of dust released by comets long ago. In fact, a meteor shower is simply the Earth crossing what was once the tail of a comet.

2. Occasionally exceptions are made to this rule, primarily in connection with the Lowell Observatory Near-Earth Object Survey (LONEOS), which has discovered a number of comets. Those discovered by Brian Skiff, an astronomer who works with the program and who already has a number of comets to his name, are called "Skiff," while others are called simply "LONEOS."

3. Writer and journalist Alistair Cooke had a weekly radio program that was broadcast on the BBC as *A Letter from America* and that ran uninterrupted from March 1946 to shortly before Cooke's death, in 2004, at the age of 95. During the fifteen-minute program Cooke offered a commentary on some aspect of American life or on a current news story viewed from an American perspective. One of his letters, from the early 1980s, was devoted to a discussion of the English language in America, including the pronunciation of British as opposed to American English. Cooke called attention to the fact that—in contrast to the situation in, say, France or Spain—America lacks any sort of agency responsible for overseeing the proper use of the country's principal language. Moreover, in certain regions of America the status of English as the first language is in jeopardy—and not for the first time. As Cooke went on to point out, during the eighteenth and nineteenth centuries English was actually not the most commonly spoken language in some parts of the country. Pennsylvania, for example, came very close at one point to declaring German the official state language.

4. According to the model of the universe that had prevailed at least since the time of Pythagorus, the known planets (Mercury, Venus, Mars, Jupiter, and Saturn) were each supported on a crystal sphere, centered on the Earth. The spheres slowly rotated, carrying the planets around the Earth in a circle—the most perfect of forms. When Halley demonstrated that his comet had an elliptical orbit and that, at its farther reach, moved out three times the distance of the outermost planet, Saturn, from the Sun, the theory of the crystal spheres was revealed as untenable. After all, as it moved outward from the Sun, the comet would have crashed into the spheres and presumably either shattered them or else been stopped dead in its tracks. Moreover, thanks to Halley's calculations, the universe suddenly expanded to three times its former size. A century later, in 1781, the discovery of Uranus would underscore the impact of Halley's discovery by proving that the space beyond Saturn was not merely inhabited by a single, transient comet but in fact had a permanent resident.

5. For the past 2,500 years the period of Halley's comet has varied in cycles of about 800 to 900 years, first getting longer and then shorter again. During these cycles, the interval between the comet's appearances ranges from a maximum of over 79 years to a minimum of about 74. Over the next few orbits its period will be on the increase, having been at the shortest ever between 1835 and 1910, when the interval was little more than 74 years.

6. The date of a comet's return is officially the date of its next perihelion, the point at which it is at its closest to the Sun, rather than the date on which it first becomes visible. Since most comets are visible only for a few weeks or months, the two dates are not usually far apart. But as telescopes become more powerful and more sensitive, we are often able to locate a comet quite some time before its official return. In the case of Comet Halley's 1986 return, the comet was first spotted four years earlier, in 1982, and was still visible in 1995.

7. I hope to view the 2061 return at the ripe old age of 101, but, however reluctantly, I must accept that the spectacular appearance of the comet in 2134 will be seen only by my descendants.

8. Magnitudes are like a golfer's handicap: the lower the number, the brighter the star. As we saw in chapter 2, Ptolemy graded stars from magnitude 1 (the brightest) to magnitude 6 (the faintest), a tradition now so firmly fixed in astronomy that it will never change. However, there are a dozen stars that are brighter than magnitude +1.0—the brightest of which, Sirius, has a magnitude of −1.4. The brightest planets can also reach negative magnitude. Mars can, for example, be as much as magnitude −2.8, and Venus as much as −4.7. In short, if an object has a negative magnitude, then it is a very bright object indeed.

CHAPTER 5. Our Moon

1. Armstrong evidently intended to say "one small step for a man," but it appears that in the excitement of the moment he forgot the "a," thus leaving a generation baffled as to his exact meaning.

2. I well remember lining up as a ten-year-old with my mother to see the small amount of moon dust on display at Bristol University. My mother's reaction was somewhat less than awed, "Is that what we've been queuing for? I can get soot that looks like that out of the chimney."

3. Think of angular momentum as an ice-skater in a spin, with the Earth as the skater's body and the Moon as the arms. When the skater is first spinning, the arms are far from the body (that is, the distance between them is relatively large), and the spin is relatively slow. But as the skater gradually pulls the arms in (thereby decreasing the distance), the spin gets faster and faster until, finally, when the skater is completely hunched up, the spin is so fast that one cannot see the details of the skater's body. In the same way, as the body of the Earth pulls the Moon in, the spin has to get faster and faster to compensate for the decreasing distance. Conversely, if the hunched-up skater's arms were reextended—as when the Moon breaks off from the Earth—the skater would spin more slowly as the distance between the skater's body and arms increased. Or imagine two skaters locked together in a rapid spin. If one of them then throws the other (increasing their separation), each of them will almost completely cease spinning as a result.

4. The title of the Fox News program was "Conspiracy Theory: Did We Land on the Moon?" A 1999 Gallup poll, however, puts the figure at about 6 percent—although that's still something

like 15 million people. According to one of the more inventive theories, Stanley Kubrick was approached by officials from NASA with a proposal that he direct staged versions of the Apollo XI and Apollo XII moon landings. This he agreed to do, working from a soundstage in Huntsville, Alabama. He refused to direct the Apollo XIII landing, however, after NASA rejected his script, in which the mission was portrayed as a failure—Kubrick being of the opinion that the public would begin to grow bored by an unbroken string of successes. (NASA later changed its mind, coming round to Kubrick's view, but they were obliged to hire a different director for the highly suspenseful Apollo XIII episode.) According to an article in the September 13, 2002, edition of the *Guardian*, a man named Bart Sibrel claims to have evidence proving that the Apollo XI astronauts never left Earth orbit. Instead, they cleverly taped a transparency of the Earth as seen from space to one of the windows of their space capsule, which they then used to produce footage allegedly shot from the Moon. In September 2002, Sibrel confronted a 72-year-old Buzz Aldrin in a Beverly Hills hotel, demanding that Aldrin take an oath on the Bible to the effect that he truly had walked on the Moon, a demand that Aldrin answered by punching Sibrel in the nose.

The theory that the moon landings were faked has proved so popular that in April 2002 the BBC dedicated a program in its *Sky at Night* series to the question. The program's host, astronomer Sir Patrick Moore, and his guest Douglas Arnold—a photographic expert who worked for Kodak during the Apollo program, supplying photographic materials to NASA—together examined and debunked some of the alleged evidence that the astronauts never landed on the Moon at all. For example, proponents of the hoax theory argue that if Neil Armstrong was the first to set foot on the Moon, then how could he be filmed as he climbed down the ladder onto the surface? Or, people ask, why are there no stars visible against the black sky in the photographs? (The explanations are, first, that there was a television camera attached to the leg of the lunar module, and, second, that the speed of the film used for the photographs—160 ISO, shot at f/11 for 1/125 of a second—was far too slow to register stars.) In addition, various other classic items of "proof" that the moon landings were fabricated—such as the letter *C* that some people have insisted they can see carved in one of the would-be moon rocks—were given a critical appraisal. The program, which is well worth a look, can be viewed online at www.bbc.co.uk/science/space/spaceguide/skyatnight/ proginfo.shtml (go to April 2002). The *Guardian* article can be found at www.guardian.co.uk/ g2/story/0,3604,791222,00.html, and the remarkable ideas of Great North Wind at www .galactic-guide.com/articles/8S12.html.

5. Much speculation has gone on as to why NASA launched Apollo VIII just before Christmas, rather than giving the overworked staff a holiday and the chance to spend a couple of days with their families. As it turns out, the CIA had information that a Soviet Zond rocket was waiting on the launching pad for a manned circumlunar flight. The liftoff of Apollo VIII was thus moved up in an effort to beat the launch of the Soviet Zond. But this was the Soviet flight that would have included cosmonaut Belyayev—the one that never took place. Had the Soviets possessed the technological capacity to send more astronauts to the Moon than the Americans could, they probably would have gone ahead with their mission, but in fact they were able to send only a single cosmonaut. After their triumphant successes during the early years of the space race, when

the Soviet program was always the first and the biggest, to reveal to the world that at this point they were unable to do as much as the Americans—and, what is more, could not do it as quickly— would have been a propaganda disaster. After the success of Apollo VIII, the Zond rocket was adapted to return samples automatically from the Moon and to land the Lunakhod car that drove around its surface. The mysterious Luna 15 probe that crashed into the Moon's surface during the flight of Apollo XI was a final Soviet attempt to get samples of moon rock back to Earth first by using a hastily prepared automated probe. Had Apollo VIII failed, however, and the American moon program suffered delay, the Soviets would have been under less pressure, and it is quite likely that they would have solved their technical problems and sent a man to the Moon first.

6. Consider just one example of the space program's contributions to electronics. If only one extra drop of solder had been put on each one of the millions of electrical connections in the Saturn V rocket, the extra solder alone would have weighed as much as the entire rocket. This was clearly a very big incentive for research aimed at developing compact circuitry and, eventually, the microelectronics that are fundamental to almost every device we now use at home or at work. Similarly, in the domain of materials science, when it takes 3,000 tonnes of rocket (consisting mainly of fuel) to launch 44 tonnes of mass into lunar orbit in order to put just 17 tonnes down on the lunar surface, it is essential to develop very strong, lightweight materials. In round figures one can say that it required 200 kilograms of fuel to put 1 kilogram of mass on the surface of the Moon and then return it to Earth, although this is in fact quite an impressive ratio. By way of contrast, using the technology available in, say, 1940, before the V2 and large, liquid-fueled rockets entered the picture, it would have taken around 1 million kilograms of fuel to put 1 kilogram on the Moon's surface. It can also be argued that had it not been for the Apollo missions the computer revolution, although inevitable in the long run, would have been greatly delayed. It is impossible to know how many billions of dollars it would have cost us to wait years longer for the advent of pocket calculators, desktop computers, portable computers, microchips, and so on.

7. On January 15, 2004, President George W. Bush announced a new initiative to return to the Moon and then send astronauts to Mars. President Bush requested an extra $1 billion for NASA, with a target of returning to the Moon no later than 2020, and possibly as early as 2015— in other words, before the fiftieth anniversary of Apollo XI. Further details of his proposals can be found in a BBC news report at http://news.bbc.co.uk/1/hi/sci/tech/3395165.stm (and see also the afterword in this volume). It is a genuinely frightening thought that early in the twenty-first century staging a manned mission to the Moon within the space of fifteen years is, in the opinion of many experts, likely to be extremely difficult, perhaps even impossible, without radical changes at NASA. In 1961, before the United States had even launched an astronaut into space, NASA was able to engineer a moon landing in just eight years, from a standing start.

8. The first space tourist was Dennis Tito, a Californian multimillionaire, who paid $20 million to spend six days on the International Space Station in late April and early May of 2001. He was followed a year later by the South African entrepreneur Mark Shuttleworth, also a multimillionaire, who spent ten days in space. New Jersey scientist and businessman Gregory Olsen had been scheduled to become the third space tourist in April 2005, once again spending $20 mil-

lion for the privilege. All three flights have been or will be in Russian Soyuz rockets, with the guests housed in the Russian crew quarters at the space station. NASA has objected very strongly to having such visitors aboard the International Space Station, but to many romantics, such as myself, the advent of the space tourist means that the space age has truly begun—although, sad to say, almost certainly too late for me to satisfy my greatest ambition and travel into space.

There are those who have inquired about the crew manifest of certain of the space shuttle flights, asking whether some of the people on board might not more properly have been called space tourists, given that their exact role as "mission specialists" was somewhat vaguely defined. Perhaps the most tactful answer is to say that, to date, only two people have actually paid out of their own pockets for the privilege of a ride into space.

CHAPTER 6. Is There Life on Mars?

1. On August 28, 2003, Mars was closer than it has been at any time in the last 58,000 years. Even so, through a telescope it was only as large as a crater on the Moon perhaps 100 kilometers in diameter.

2. Lowell also depicted canals on other planets, including Venus, which is in fact almost featureless when seen through a telescope. To be fair, though, recent research with images from space probes has revealed occasional, very faint markings in the clouds of Venus that do look somewhat like Lowell's canals, although this discovery is unlikely to rehabilitate Lowell's image as an observer of the planets.

3. At 4.2 meters in diameter, I suspect that this is the largest telescope ever to have been used with an eyepiece. Most modern telescopes of any considerable size are designed to be used only with electronic detectors controlled by computers and therefore have no provision for an eyepiece to be fitted anywhere. In fact, for at least fifty years now, almost no planetary observations have been carried out visually on large professional telescopes. My visual inspection of Mars took place on a staff training night, when the telescope was out of service for normal use. The sight of the planet was quite stunning through the eyepiece.

4. Images from Mars Pathfinder, which landed on Mars in 1998, show that some of the rocks at the landing site have a coloration strikingly similar to that of the blue-green markings seen from Earth. These markings probably represent areas of the planet where there is much less dust and more native rock on the surface.

5. Mars 2 impacted a probe displaying a Soviet pennant in the southern hemisphere. In the midst of the still raging dust storm, Mars 3 soft-landed a probe on the opposite side of the southern hemisphere. This lander operated for only twenty seconds, however. It had begun to send an image, but the transmission was lost before the picture showed anything. In both cases, the orbiter did return some useful data, although the images were far inferior in quality to those of Mariner IX.

6. It was earlier thought that Olympus Mons could be as much as 26 kilometers in height, but Mars Global Explorer Laser Altimeter (MOLA) profiles have revealed it to be about 5 kilometers shorter.

7. Quoted in J. Kelly Beatty and Andrew Chaiken, eds., *The New Solar System*, 3rd ed. (Cambridge, Mass.: Sky Publishing, 1990), p. 278. Soffen's well-known remark originally appeared in the *New Scientist*.

8. Quoted in Beatty and Chaiken, eds., *The New Solar System*, p. 279.

CHAPTER 7. Pluto

1. In "Account of a Comet," a paper Herschel read on April 26, 1781, before the Royal Society—an organization that embraced all scientific disciplines—and that appeared in the society's ongoing record of its proceedings, *Philosophical Transactions* (1781, doc. 32). The small star in the quartile between Auriga and Gemini was 132 Tauri, and H Geminorum is now known as 1 Geminorum.

2. All the players in our story had brilliant future careers in astronomy apart from Adams himself, who, after the discovery of Neptune and the violent furor it caused, all but disappears from the history of astronomy. He was a shy and sensitive man, and one wonders whether he simply decided that enough was enough. He even later declined to accept a knighthood from Queen Victoria, although he was twice president of the Royal Astronomical Society. In 1853, at the age of only 34, he published a seminal paper on the movement of the Moon, but this appears to have been his final major contribution to astronomical research. Adams died, aged 73, in 1892.

3. Until the twentieth century the two most sought-after posts in astronomy were the directorships of the Royal Greenwich Observatory (RGO) and that of the Paris Observatory, arguably the two most important astronomical centers in the world. The Paris Observatory still exists in the form of the Paris-Meudon Institute and Observatory, but, sadly, the RGO was finally closed in the 1990s after a short-lived and ill-fated move to Cambridge. The name of Greenwich lives on in the form of the old observatory in Greenwich Park, the site of the RGO until 1950, which is now somewhat confusingly named the Royal Observatory Greenwich.

4. Before the Celestial Police began their search, Giuseppi Piazzi, the director of the Palermo Observatory, had already located the first, and largest, of the asteroids, Ceres. His discovery was made on January 1, 1801—so, at least astronomically speaking, the nineteenth century started out with a spectacular bang.

5. As it turns out, Galileo himself had observed Neptune on three occasions, when the planet was close to Jupiter. Sketches he made on December 28, 1612, and on January 2 and 28, 1613, clearly show Neptune. But it was only in 1980 that these observations were finally recognized for what they were.

6. This was one of those lunatic legal cases where the only victors in the end are the lawyers. Neither Lowell's widow nor his observatory ultimately received the money, which was spent almost in its entirety on legal fees. When the case was finally settled in 1927, even $10,000 for a new telescope was far more than the observatory could afford.

7. Clyde Tombaugh and Patrick Moore, *Out of the Darkness* (Harrisburg, Pa.: Stackpole Books, 1980), p. 118.

8. Ibid., p. 127.

9. The naming of asteroids and TNOs is a rather complex affair. When an asteroid or a TNO is first discovered, it is given a local designation, which is then reported to the Minor Planet Center. For example, when the first asteroid was discovered at Teide Observatory in Tenerife, it was designated "Teide5" (since "Teide1" through "Teide4" turned out to be objects already catalogued). Then, once the Minor Planet Center had checked the observations and verified that this was indeed a previously unidentified object, Teide5 received the provisional official designation "1998 EO$_4$"—"1998" for the year of its discovery and "E" (the fifth letter of the alphabet) because it was discovered on March 6, which falls in the fifth half-month of the year (March 1 to March 15). As for the O$_4$, within each half-month objects are labeled A, B, C, etc., in order of their discovery. When you get to Z (having skipped over "I," which is too easily confused with "1"), you return to the start of the alphabet and add a subscript "1"—A$_1$, B$_1$, C$_1$, etc.—until you reach Z$_1$, at which point you begin the alphabet again with a subscript "2," and so on.

When an object has been observed a sufficient number of times over a sufficient number of years that its orbit is very well known, it is awarded an official number. Although 1998 EO$_4$ has now been observed in five separate years, it has not yet been observed frequently enough for the Minor Planet Center to consider it worthy of a number—although it is likely to receive one if it is observed again next year. In contrast, another Teide Observatory asteroid, 1998 GE$_1$, was observed extensively enough over several years to be given a number in the year 2003. This object is now asteroid 44103. Once the asteroid is numbered, its discoverer may choose to give it a name. The proposed name is then submitted to the Small Bodies Naming Committee of the IAU, which votes on it. The name is accepted only if no one on the committee objects to it. Hence "Teide13" became first "1998 GE$_1$" and then "(44103) 1998 GE$_1$" and is now known as "(44103) Aldana." Another object, discovered by amateur astronomers in Majorca, has become "(20141) Markidger" (according to IAU convention, whereby doubled letters are disallowed).

10. On finishing his extended search in 1943, Tombaugh was given a leave of absence from Lowell Observatory so that he could teach navigation in the armed forces. In 1945, after the war ended, Tombaugh returned to Lowell and was promptly sacked. The reason given was cost cutting, although this convinces no one. Some people suggest that the presence on the staff of an astronomer who had little formal education but who was a world-famous planet discoverer would have created too much internal tension. But Tombaugh had worked happily at Lowell, continuing the search for trans-Neptunian planets, for thirteen years after discovering Pluto, so this explanation does not ring true either. He eventually took a position at the White Sands Proving Grounds and later taught at the University of New Mexico.

11. The less expensive Voyager mission was originally intended to visit only Jupiter and Saturn. Voyager 1 could theoretically have been redirected to take in Pluto as well, but it was instead programmed to encounter Saturn's large moon Titan, making a visit to Pluto impossible.

12. Malcolm W. Browne, *New York Times*, February 9, 1999. In the end, another asteroid was crowned number 10,000—although a serious suggestion was made that 10,000 be left open, with the list jumping straight from 9,999 to 10,001. It is a reflection of the frenetic rate of asteroid discoveries that as of February 1, 2004—not quite five years after asteroid 10,000 was numbered

on March 2, 1999—there were 73,636 numbered asteroids, and this number has increased to 90,671 from the time I submitted the final manuscript of this book to the completion of copy editing, in mid-September. It thus seems certain that by the end of March 2005 we will have reached asteroid number 100,000.

CHAPTER 8. How Astronomers Learn without Going Anywhere

1. It was not until around 1850—well after Herschel's time—that photography began to be adapted for use in astronomy. The early photographic plates were so insensitive, though, and the necessary exposures so long that satisfactory images began to be obtained only in the 1880s (and in 1892 an asteroid was for the first time discovered photographically). In theory, because the telescope at Birr Castle was completed in 1845, Lord Rosse could at some point have attached a camera to it, but in fact his observations take the standard form of drawings. Then again, Ireland is not famed for its balmy weather and cloudless skies. As someone once aptly commented, the miraculous thing was not *what* Lord Rosse saw with his telescope but rather that he saw anything at all.

2. As a construction site the Moon offers a number of advantages. It has no atmosphere that might blur photographic images and no weather that might damage the telescope, and its gravity is far lower than the Earth's, which makes it easier to build very large structures there. Moreover, light pollution would not be a problem since the Moon is notably lacking in urban sprawl (at least so far). However, before we can build telescopes the size of OWL, we will require faster and more sophisticated computers as well as more experience with the construction of huge instruments that demand micrometric precision. For example, the ring, 17 meters in diameter, on which the 11.4-meter GTC will move has twice had to be ripped up and relaid because its placement was off by a single millimeter—and the pieces of the mirror must be positioned with an accuracy of better than a *ten-thousandth* of a millimeter. Whereas the mirror of the GTC is made up of a mere 38 hexagonal segments, the mirror of a 100-meter telescope will consist of somewhere between 600 and 1,000 smaller individual mirrors, each one of which will need to be put into position, and then held in place as the telescope moves, with the same amazing degree of precision.

3. The discovery of Io's volcanoes came down to a piece of eagle-eyed work by an engineer at the Jet Propulsion Laboratory named Linda Morabito, who was measuring images of the satellites in order to calculate the precise position and trajectory of the probe and thus improve the tracking data for the spacecraft. On one frame she noticed a small crescent shape by the limb of Io. Thinking this must be another of Jupiter's satellites moving behind Io, she checked to see which one it was and discovered that there was no other satellite there. When the images were examined in more detail, it turned out that what she had seen was a huge volcanic plume—a fountain more than 300 kilometers high—shooting out of the volcano's top.

4. It was miraculous that the Galileo managed to send anything at all back to Earth. Its big umbrella antenna jammed, refusing to open, and so the probe was limited to a low-power transmitter used with an omnidirectional antenna (like the one for your car radio) that permitted only a very slow speed of transmission—a mere 10 bytes (0.01 kilobytes) per second, or roughly five

thousand times slower than a modem, which typically offers 56 kilobytes per second. At the start, each photograph required some 30 hours to be transmitted, although improvements to the transmission software later boosted the data rate to 40 bytes (0.04 kilobytes) per second, cutting the transmission time to about 8 hours per image. For each fly-by scientists therefore faced a choice: they could either take a few photographs or else record data pertaining to one of the other experiments. But they couldn't do both because it was not possible to store and transmit that much data to Earth.

5. Oddly, despite the strange fact that the atmosphere is rotating at a far different velocity and in the opposite direction, wind speeds on the surface of the planet are rather low, evidently just a few kilometers per hour. However, what with Venus's enormous atmospheric pressure, more than ninety times that on the Earth's surface, the gentle breeze has a force equivalent to the waves from a storm at sea crashing into the shore on Earth.

6. In the Giotto images, at the probe's closest approach to Comet Halley the nucleus had slid out of the camera's field of view. The last images that captured at least part of the nucleus itself were taken about a minute before the encounter, at a distance of some 2,000 kilometers.

7. The farther from us a comet is, the fainter it appears to be, regardless of its actual degree of brightness, and the farther it is from the Sun, the less sunlight there is for it to reflect. To quantify the true brightness of a comet what we therefore do is calculate its absolute magnitude, that is, how bright the comet would be if it were simultaneously 1.0 AU from the Sun and from the Earth. In other words, to calculate how bright the comet in fact is, we must adjust for its distance from both the Sun and the Earth at the time.

CHAPTER 9. Clear and Present Danger

1. Larry Niven and Jerry Pournelle, *Lucifer's Hammer* (London: Orbit Books, 1977), p. 157.

2. Whether or not a planet has an atmosphere is all but irrelevant in the case of asteroids above a certain size. An ordinary stony asteroid that is about 50 meters or more in diameter will reach the planet's surface as if the atmosphere did not exist—and if the asteroid contains a substantial amount of iron, it need not be larger than a meter or two across to do so.

3. There are two exceptions: Io, the closest of Jupiter's four large satellites, and Europa, the second closest. Io has such an unstable surface, covered with so many extremely active volcanoes, that impact craters are wiped out almost as soon as they form. Europa has a surface of ice, apparently floating on top of an ocean hundreds of kilometers deep, on which only a few impact craters have been identified. Such craters are probably destroyed soon after forming, either because the ice breaks up for some reason or because the edges of them flatten out and flow away under their own weight. Saturn's largest satellite, Titan, has recently been mapped in part by radar from the Cassini space probe. The images show various features that are obviously impact craters.

4. Quoted in Carl Sagan, *Cosmos* (New York: Random House, 1980), p. 85. My grateful thanks to Graeme Waddington for his assistance with the interpretation of Gervase's text. Interestingly, nowhere does Gervase say that the men were monks (although Gervase was). This appears

to be a supposition on the part of Sagan, although it is a reasonable enough one. As monks, they would be literate and thus able to record any strange phenomenon that they observed.

5. The orbit of Comet Encke actually crosses the Earth's orbit twice a year—in June, on the outward leg of its orbit, after being at perihelion at approximately the same distance from the Sun as the orbit of Mercury, and in November, when the comet is inbound, before it rounds the Sun. During the month of June, instruments left on the Moon by the Apollo missions have registered the impact of what may be fragments of the comet. This material consists of dust and small boulders that broke away from the comet at some point in the past and have been held in an orbit around the Sun roughly similar to that of the comet itself. As the comet is moving away from perihelion and toward the Earth, these fragments come toward us from more or less the direction of the Sun in the sky. This produces a meteor shower in the daytime sky, a phenomenon that can be observed only by radar. Much the same thing happens in November, except that we are considerably further from the orbit of Comet Encke in November than in June, and so we encounter far fewer fragments and only a weak meteor shower.

6. Using Dance of the Planets, a computer simulation program developed by Thomas R. Ligon, I calculated that on June 18, 1178, the 1.7d Moon was 2° high with the Sun at −6° (that is, at civil twilight) and that the Moon set 1h12m after the Sun.

7. Most comets have eccentric orbits and fall in toward the Sun at high velocity. They can approach the Sun from any direction, and so a head-on collision with the Earth at an extreme velocity—more than 70 kilometers per second—is possible. In contrast, asteroids usually have fairly circular orbits, and they orbit the Sun in the same direction as the Earth. The collision of an asteroid is therefore generally the result of a tail chase, and the asteroid's velocity relative to the Earth is much lower. Think of it like an automobile accident: a rear-end collision can certainly cause injuries, but a head-on collision is very often fatal.

8. See Donald K. Yeomans, *Comets: A Chronological History of Observation, Science, Myth, and Folklore* (New York: John Wiley and Sons, 1991), pp. 186–87.

9. When a comet or asteroid has been lost and is then relocated, whether accidentally or deliberately, this is called a "recovery." A "prediscovery" image is one made when a newly discovered object is located on fairly recent photographs. For example, Comet Hale-Bopp was photographed several times before it was actually discovered and then located on these images after the discovery was announced. A "precovery" is made when an asteroid is located in a deliberate search of old archival images, sometimes taken many years beforehand. Some recently discovered asteroids have even been pinpointed on images taken at Mount Palomar in the early 1950s.

10. J. D. Giorgini et al., "Asteroid 1950 DA's Encounter with Earth in 2880: Physical Limits of Collision Probability Prediction," *Science* 296, no. 5565 (April 5, 2002), pp. 132–36.

11. In fact, at the time only two other amateur astronomers besides Ferrando had ever discovered NEAs—that is, all five of these asteroids were first identified by one of the three. Since then, amateur astronomers in Spain have located another NEA—Apollo object 2003 QA, which is nearly a kilometer in diameter. The asteroid was discovered by Salvador Sánchez and his team at the Observatorio Astronómico de Mallorca on August 16, 2003.

12. "More on the Great Impact Debate," CCNet 24/2003, 3 March 2003, at http://abob
.libs.uga.edu/bobk/ccc/cc030303.html. Also very worthwhile is the paper that Paine coauthored
with Benny Peiser (and for which he helpfully cites the Web address): "The Frequency and Consequences of Cosmic Impacts since the Demise of the Dinosaurs" (2002).

CHAPTER 10. Goldilocks and the Three Planets

1. As we saw in chapter 8, Venus rotates on its axis roughly every 243 of our days, but the
planet's rotation around the Sun shortens the apparent length of its day. That is, as Venus turns
on its axis, it is also revolving around the Sun. As the planet turns, its lighted face changes, but
because the planet is also circling around the Sun, the direction from which the Sun's light is shining also changes. The sum of the two effects is such that the time between sunrise and sunset (59
of our days) is far less than the actual length of one Venusian day.

2. Desperate attempts were made to recover contact, including one unprecedented Soviet re-
quest for assistance from England's famous radio telescope at Jodrell Bank, near Manchester. Unfortunately, though, all efforts were in vain.

CHAPTER 11. Going to the Planets?

1. Captain W. E. Johns wrote a series of children's novels about the exploration of Mars, with
such titles as *The Kings of Space* and *Return to Mars*. He is best known in England, however, for his
famous series of "Biggles" books, about a daring pilot, James Bigglesworth, and his adventures as
a fighter ace in the First World War and, later, in the Special Air Police. Between novels and col-
lections of short stories, these books ran to over a hundred volumes, published between 1932 and
1970, and were all but required reading for children, especially in the war years of the 1930s and
1940s.

2. *Battlestar Galactica* was originally an American television series that ran for only twenty-four
episodes, between September 1978 and April 1979. The series began from the premise that the
human race had been exiled from its various colonies around the Galaxy after losing a war with
the evil robot Cylons. Led by the giant starship *Battlestar Galactica*, the survivors try desperately to
find the lost planet Earth so that they can resettle it. The series spawned a 1978 film of the same
title, which appeared shortly after *Star Wars* and had a number of elements in common with it, leading to accusations that the makers of *Battlestar Galactica* were trying to cash in on the success of
the George Lucas film.

Arthur C. Clarke has remarked that, although entertaining, *Star Trek* is not really science fiction but fantasy. For example, in the real world, we will never exceed the speed of light. "No Warp
Six will ever get you from one episode to another in time for next week's installment," he comments
in the author's note that introduces his novel *The Songs of Distant Earth* (London: Grafton Books,
1986), on p. xiii.

3. One would hope that by the time the human race is able to provide clean, low-cost energy
to the whole world from the Moon, our planet will have settled its differences so that the potential temptation to use such a powerful microwave beam for military purposes can be safely set aside.

That said, in his children's novel *Terror by Satellite* (New York: Faber & Faber, 1980 [1964]), Hugh Walters presents a disturbing vision of the wrongful use of a satellite in space that was intended simply for generating power from solar energy and beaming it to Earth but was instead made to serve as a weapon of terror and blackmail.

4. Prior to the Clementine and Lunar Prospector probes, a number of science fiction authors—for example, Isaac Asimov in *The Gods Themselves* (London: Victor Gollancz, 1972)—had suggested that ice might exist in the lunar interior or in the form of frost inside caves or on mountaintops. A few writers even imagined that there might be enough ice to be worth mining, but nobody seems to have anticipated the truly enormous quantities of ice that the findings from the Lunar Prospector mission indicate may exist. Not everyone is convinced by the sum of the evidence, however. Skeptics include the astronaut Dr. Harrison Schmidt, the only geologist to have set foot on the Moon, as well as the renowned British astronomer Sir Patrick Moore.

5. Of the 3,000 tonnes that the Saturn V rocket weighed when launched, 150 tonnes of payload reached Earth orbit, of which a total of 54 tonnes entered orbit around the Moon (the command and service module and the lunar module itself). For the Apollo XVII mission, which had the heaviest total payload, there were 16.6 tonnes of fully fueled lunar module, of which just 6.7 tonnes—slightly more than 0.2 percent of the total weight of the Saturn V—arrived on the lunar surface.

6. Sir Richard Branson will join with the two founders of Mojave Aerospace Ventures, aircraft designer Burt Rutan and Paul G. Allen, cofounder of Microsoft, to form a company called Virgin Galactic. The plane for which Virgin has contracted will be a larger version of Mojave Aerospace's SpaceShipOne, which in June 2004 became the first privately built rocket ship to enter space. Out of the twenty-plus spacecraft entered in the competition, SpaceShipOne has now won the $10-million Ansari X-Prize, which was to be awarded to the company that produces the first spacecraft designed and constructed without government participation that is able to carry three or more persons at least 100 kilometers into space and then to do so a second time, using the same spacecraft, within two weeks. Virgin's plane—the first of which is to be christened the V.S.S. Enterprise—will carry five passengers. For more information, see the BBC News World Edition at http://news.bbc.co.uk/2/hi/science/nature/3693020.stm, http://news.bbc.co.uk/1/hi/sci/tech/3693518.stm, and http://news.bbc.co.uk/1/hi/sci/tech/3676312.stm. See also Heather Timmons, "Virgin to Offer Space Flights (Even, Sort of, at Discount)," *New York Times*, World Business section, September 28, 2004.

7. Two such novels that I particularly recall are *Footfall*, by Larry Niven and Jerry Pournelle (New York: Ballantine Books, and London: Victor Gollancz, 1985), and *Passage to Pluto*, by Hugh Walters (New York: E. P. Dutton, 1975).

8. "Buzz Aldrin: Down to Earth," *Psychology Today*, May–June 2001. The interview, which is quite a tribute to the human spirit, is available online at http://health.yahoo.com/health/centers/depression/2195.

9. Most commonly, spacecraft are launched into an orbit close to the equator (hence the location of the John F. Kennedy Space Center at Cape Canaveral, on the east coast of Florida) and

in the direction of the Earth's rotation around its axis, since this gives the spacecraft a boost of 900 miles per hour—the speed at which the Earth's surface turns. The further away from the equator a launch site is, the less a spacecraft will profit from this boost, which translates into more fuel and less payload. An equatorial orbit also has the advantage of allowing the spacecraft to pass over the same points on the Earth on every circuit, which among other things makes it easier for astronauts to stay in touch with tracking stations. Unfortunately, though, even from a high geostationary orbit, less than half of the Earth's surface is visible, sometimes only from a great distance, and much of it only at a slant. In order to be able to observe the entire surface of our planet, a spacecraft must be launched into a polar orbit, that is, into an orbit that takes it over the Earth's poles rather than around the equator. As the spacecraft orbits, the Earth rotates beneath it, bringing every point on the planet into view once a day. This is the orbit that the famous "Keyhole" spy satellites use.

The only difficulty (at least for the United States) is that if a spacecraft is launched into a polar orbit from Florida, it will pass over densely populated areas during liftoff, rather than over the sea. Such a route clearly puts human lives at risk should an accident occur—as indeed happened with Mariner VIII, which lifted off only to fall back again to the Earth, and with the explosion aboard the Space Shuttle Challenger. Because they were launched on equatorial trajectories, both spacecraft fell into the ocean, just as they were supposed to in the event of an emergency. Had they been aimed at a polar orbit, however, the consequences could easily have been catastrophic. Polar launches are thus made instead from Vandenberg Air Force Base, in California, roughly halfway between San Luis Obispo and Santa Barbara. Being on the West Coast close to several major cities, Vandenberg is not suitable for equatorial launches other than those aimed at "wrong way" orbits, against the direction of the Earth's rotation, but its location does allow spacecraft to take off over the Pacific Ocean on a polar trajectory.

CHAPTER 12. Are We Stardust?

1. This was quite a lesson about the gap between television—even an otherwise factual documentary—and reality. For example, the first day of filming was brilliantly sunny and warm and invited work in shirtsleeves. But, of course, we had to be dressed identically on the second day to make the footage match. Predictably, the weather changed and became windy, foggy, and bitterly cold. After filming each scene, in which naturally we had to look as if everything was normal, the only thing on our minds was to put on as much warm clothing as we could as fast as possible!

2. Even among the 92 "naturally" occurring elements, some are actually so unstable that they are not found anywhere on Earth. The classic example is technetium (atomic number 43), which, as its name suggests, can only be manufactured in laboratories. (Much to the surprise of scientists, however, technetium has been discovered in a few red giant stars, where it evidently forms deep within the star's interior—and in sufficiently large amounts that some of it is able to survive long enough to reach the star's outer shell, where it can be detected.) In addition, a few of the elements heavier than uranium, such as plutonium (atomic number 94), are found naturally in uranium-bearing rocks. For the most part, though, the elements beyond uranium are too unstable to occur naturally. Scientists have been able to create them artificially, although in some cases for only a very

brief period, given that they are exceedingly unstable. Owing to the delicacy and complexity of these experiments, the results are occasionally uncertain. In 1999 scientists from the Lawrence Berkeley Laboratory claimed to have detected element 118 in experiments involving the collision of atoms of lead and krypton. But no one has since been able to detect the element, and the announcement has now been retracted. To date, the Los Alamos National Laboratory lists twenty-three elements beyond uranium as having been observed or detected, although two of them—elements 116 and 118—remain unconfirmed. The largest of those that have been confirmed are elements 112 and 114. (More information can be found at http://pearl1.lanl.gov/periodic/default .htm.)

3. Every kind of elementary particle has its corresponding antiparticle. Electrons and antielectrons (positrons) and likewise neutrinos and antineutrinos belong to a group of elementary particles called leptons, which are fundamentally different from ordinary particles such as protons and neutrons (known to physicists as baryons). The laws of particle physics say that the balance of leptons and antileptons cannot change—that is, the number of leptons and antileptons must always sum to zero. As we have seen, when a proton becomes a neutron, something must be formed to carry off the proton's positive charge, and this something is the positively charged antielectron (or positron). But by forming this antielectron, we have created one lepton, which in this case happens to be an antilepton. So, because we have formed an antilepton—a "minus" lepton— we must also form a "plus" lepton so that the sum of the two is effectively zero leptons. Given that the electrical charges have already been accounted for, however, this lepton cannot be a charged particle. To allow for this situation, nature cleverly created the neutrino. A neutrino has a lepton number—that is, it counts as one lepton, just as its antiparticle, the antineutrino, counts as one antilepton—but it has no charge and virtually no mass, making it almost impossible to detect.

4. In a Type II supernova, the nucleus of the star—the remaining 10 percent, which is composed largely of iron—is crushed by the tremendous force of the blast. As a result, protons and electrons are forced together and combine to form neutrons, producing what is known as a neutron star. Astronomers are still arguing about precisely what Type I supernovas are. The most popular theory is that they involve double stars in which one of the stars is a white dwarf, similar in mass to the Sun, and the other a large star. Material from the large star, most of which is "unburnt" hydrogen from the outer layers, keeps falling onto the white dwarf, and the mass of the dwarf star progressively increases as a result. Up to a point, a force known as "degeneracy pressure"—essentially the resistance of the electrons in the star to being crushed too closely together— is able to counterbalance the force of gravity. But once the mass of the white dwarf exceeds 1.4 times the mass of the Sun, the force of gravity becomes so great that it overwhelms the degeneracy pressure, at which point the star collapses. All the hydrogen that has fallen on the white dwarf suddenly combines to form helium in what is termed a "nuclear deflagration"—an uncontrolled nuclear explosion on a huge scale—and the star blows itself apart. This is the standard explanation of what happens in a Type I supernova. An alternative, and more recent, theory, suggests that Type I supernovas might be produced by the merging of the two components of a binary white dwarf. In this case, two individual stars forming a binary system both turn into white dwarf stars

(although doubtless not at the same time). Then, over the course of many millions of years, the two white dwarfs spiral into each other and merge, again with a resulting nuclear deflagration.

5. This may seem counterintuitive, but the larger a star, the shorter its lifetime—because the larger it is, the more rapidly it uses up its fuel. Our Sun will last for about 10 billion years, whereas a star one hundred times as massive as the Sun will last for only a few million years—and one a tenth as massive as the Sun may be around for hundreds of billions of years.

6. Because these small white grains are rich in calcium and aluminum, they go by the name of CAIs: Calcium- and Aluminum-rich Inclusions.

Index

atom bomb. *See* nuclear weapons
Avebury (England), 9, 10

radiocarbon dating, 14–15

research, scientific: and Apollo missions, 97–100; applied *vs.* theoretical, 1–3; and spaceflight, 240, 245

Romans, 37–38, 53–56

Rosse, Lord, 161, 163, 283n1

Royal Greenwich Observatory (RGO), 139, 163, 281n3

Russian space program, 238, 280n5, 280n8, 286n2 (chap. 10); *vs.* American, 87, 90, 97, 98, 240, 245, 278n5; cosmonauts in, 267, 270, 278n5; tragedies in, 88, 270

Sagan, Carl, 115, 116, 188, 284n4

saros cycle, 19–20, 39

satellites, military, 243–44, 288n9

satellites, planetary, 90–92, 156, 157, 186, 241, 262, 284n3; asteroids as, 92, 95; naming of, 274n1; and Pluto, 91, 150, 153–54, 155; volcanoes on, 168–71

Saturn, 38, 39, 135, 262, 276n4; and Comet Halley, 78, 79; missions to, 153, 237, 238, 240, 282n11; rings of, 240–41; satellites of, 90, 91, 95, 156, 186, 284n3; and star of Bethlehem, 63–64, 68

Saturn V rocket, 202, 236–38, 243, 269, 279n6, 287n5; fuel for, 232–33

Schiaparelli, Giovanni Virginio, 109–10

science fiction, 213, 228, 238, 240, 287n4; and colonization of solar system, 182, 224–25, 243; disaster scenarios in, 183–85; *vs.* fantasy, 224–25; on Mars, 107, 111, 242. *See also particular authors*

silicon, 228, 251, 254–55, 256, 260

Sky at Night (TV program), 49–50, 278n4

Slipher, Earl Carl, 145, 146, 147

Slipher, Vesto, 145, 146, 147, 150

Small Bodies Naming Committee (IAU), 73, 75, 282n9

solar system, 156; colonization of, 182, 224–25, 231–32, 239–42, 243; dangers in, 185–86, 190; exploration of, 87–180; formation of, 219, 260–63; Ptolemaic, 30–33, 40

Soviet Union. *See* Russian space program

Soyuz spacecraft, 244, 270, 280n8

spaceflight: abandonment of, 244–45; benefits of, 1–3, 99–100, 234–35, 245, 279n6; commercial, 103–4, 236, 243, 245; economics of, 97, 227–29, 233, 234–35; fuel for, 232–35; manned, 236, 242–45, 265–71; medical issues of, 236–37; from Moon, 232–35; risks of, 88, 270–71; technology for, 235–39, 240

space probes, 133, 168, 194, 203; Beagle 2, 126, 127; Clementine, 102, 230; Galileo, 169, 170, 226, 283n4; Giotto, 84, 175, 176, 284n6; Lunar Prospector, 102, 103, 230–32, 239; Magellan, 173, 174; manned *vs.* unmanned, 88, 225–26, 240; to Mars, 113–14, 126, 127, 226; Mars Express, 126, 127; Pioneer Venus, 173, 174, 214; Surveyor, 125, 190; Venera, 160, 214, 215; Viking, 166. *See also* Mariner space probes

space shuttle (Space Transportation System; STS), 233, 236, 243–44, 266–68. *See also* Challenger; Columbia

space stations, 234. *See also* International Space Station

spectroscopy, 117–18, 127, 162, 163

stars, 66, 67, 224, 260; catalogues of, 31, 40, 46; composition of, 160, 254–55; in constellations, 30–40, 41, 45–46; and human race, 182, 249, 262; magnitudes of, 31, 274n2, 277n8; mapping of, 27–33; naming of, 26–48; numbers for, 40–41; peculiar names for, 41–43; purchase of, 26, 46; reactions within, 254–55. *See also* Bethlehem, star of

Stonehenge, 4, 9–25; as astronomical computer, 18–20; construction of, 13–18, 29; dating of, 14–15, 18; holes at, 15–16, 18, 19, 20, 21, 29; modern appearance of, 10–11; and Moon, 16, 19; purpose of, 11, 18–23; reconstruction of, 11, 14; stones in, 11, 12, 16, 17, 18, 19, 21, 30

Sun, 91, 92, 93, 173, 188, 220, 224, 286n1 (chap. 10); and asteroids, 193, 202–3, 204; collapse of, 254, 290n5; composition of, 160, 251, 252, 253–54; distance of planets from, 152; in early astronomy, 29, 31, 38; electricity from, 228–29; and formation of solar system, 261, 262; hypothetical companion of, 192–93; reactions within, 252–54; study of, 162, 165, 166, 167; worship of, 21–22

supernovas, 62, 67, 224, 255–63; collision of, 258–60; composition of, 256, 257, 260, 262–63, 289n4; and formation of solar system, 260, 262–63; nuclear reactions in, 256; Sn1987a, 255–56

technology, 198; as benefit of space exploration, 1–3, 99–100, 234–35, 245, 279n6; and changing course of asteroids, 202–3; and ice on Moon, 231–32; medical, 99; military, 230, 243–44, 286n3, 288n9; for space exploration, 227, 235–39, 240, 279n6

telescopes, 43, 139, 141, 145, 163–65, 175; CELT, 164; Gemini, 126, 164, 177, 178; Herschel, 112, 134–35, 161, 163, 164; infrared, 126, 178; Keck, 170, 177; LINEAR, 74–75, 194, 204; and Mars, 108, 109, 126; OWL, 170, 283n2; radio, 104, 165, 172; Spacewatch, 200, 201; Subaru, 164, 177; visual observation with, 178, 280n3; X-ray, 168. *See also* Hubble Space Telescope

Titan, 91, 156, 240, 282n11, 284n3

Tombaugh, Clyde, 132, 145–47, 148, 150–52, 155, 158, 282n10

tourism, space, 235, 242, 279n8, 287n6; on Moon, 103–4

trans-Neptunian objects (TNOs), 148, 150–51, 155–58, 282n9

tsunamis, 192, 206–7

Tunguska event (1908), 187–88, 196, 205

Uranus, 45, 262, 276n4; and Comet Halley, 79, 80; discovery of, 134–36; missions to, 153, 240; and Neptune, 138, 140, 141–42, 143, 152; orbit of, 135, 136, 142; and Pluto, 132, 144, 148, 154; satellites of, 90, 91, 95, 156, 186; and star of Bethlehem, 59–60

Venus, 4, 68, 88, 90, 95, 160, 166, 172–74, 185, 240, 276n4, 277n8, 286n1 (chap. 10); atmosphere of, 167, 214–15, 241, 284n5; canals on, 111, 280n2; *vs.* Earth, 210–13, 217; greenhouse effect on, 219–20; and star of Bethlehem, 49, 59, 63; temperature of, 213–15; volcanoes on, 174, 178; water on, 213–15, 219–20, 241

Viking landers, 115–18, 123, 126, 127, 216–17

volcanoes: on Io, 168–71, 172, 178, 283n3; life in, 128–29; on Mars, 113–14, 123, 127–28; on Moon, 100; temperature of, 168–71; on Venus, 174, 178

Voyager missions, 153, 169, 170, 186, 282n11

water, 218, 237, 251, 262; on Mars, 114, 118, 120, 122–26, 128, 215–17, 241, 242; on Moon, 100, 229–32; on Venus, 213–15, 219–20, 241

Whipple, Fred, 148, 175, 213, 214, 219

zodiac, 29, 31–33, 38–39, 57, 64

Zoroastrianism, 58–59

Acronyms used in the listing below are as follows:

ASTER: Advanced Spaceborne Thermal Emission and Reflection Radiometer
AURA: Association of Universities for Research in Astronomy
ERSDAC: Earth Remote Sensing Data Analysis Center
ESA: European Space Agency
ESO: European Southern Observatory
GSFC: Goddard Space Flight Center
JAROS: Japan Observation System Organization
JPL: Jet Propulsion Laboratory
LaRC: Langley Research Center
METI: Japan's Ministry of Economy, Trade, and Industry
MISR: Multiangle Imaging Spectro-Radiometer
MIT: Massachusetts Institute of Technology
NASA: National Aeronautics and Space Administration
NIMA: National Imagery and Mapping Agency
STScI: Space Telescope Science Institute
USGS: United States Geological Survey
2MASS: Two Micron All Sky Survey

Gallery following page 74

Page G1. Mary Ann Sullivan, Bluffton University
Page G2. Mary Ann Sullivan, Bluffton University
Page G3. Mary Ann Sullivan, Bluffton University
Page G4. © Copyright The Trustees of The British Museum
Page G5. JPL
Page G6. From Mark Kidger, *The Star of Bethlehem*, © 1999 Princeton University Press. Reprinted by permission of Princeton University Press.
Page G7. From Mark Kidger, *The Star of Bethlehem*, © 1999 Princeton University Press. Reprinted by permission of Princeton University Press.
Page G8. From Mark Kidger, *The Star of Bethlehem*, © 1999 Princeton University Press. Reprinted by permission of Princeton University Press.
Page G9. NASA, JPL, and Malin Space Science Systems
Page G10. NASA and JPL
Page G11. NASA, JPL, and University of Arizona
Page G12. JPL
Page G13. NASA and JPL
Page G14. NASA, JPL, and Space Science Institute
Page G15. NASA, ESA, and Erich Karkoschka (University of Arizona)
Page G16. JPL
Page G17. NASA

Page G18. NASA and JPL
Page G19. NASA and JPL
Page G20. NASA
Page G21. NASA and JPL
Page G22. JPL
Page G23. NASA
Page G24. NASA, JPL, and Malin Space Science Systems
Page G25. NASA
Page G26. NASA and JPL
Page G27. NASA and JPL
Page G28. NASA, JPL, Cornell, and USGS
Page G29. NASA, JPL, and Malin Space Science Systems
Page G30. ESA
Page G31. JPL
Page G32. U.S. Naval Observatory

Gallery following page 170

Page G1. NASA
Page G2. NASA and JPL
Page G3. NASA and JPL
Page G4. NASA
Page G5. Alan Stern (Southwest Research Institute), Marc Buie (Lowell Observatory), NASA, and ESA
Page G6. JPL
Page G7. NASA and JPL
Page G8. NASA and JPL
Page G9. JPL
Page G10. © 1986 MPS; special thanks to Dr. H. U. Keller for supplying the image
Page G11. NASA and JPL.

Page G12. *Top:* Calvin J. Hamilton. *Bottom:* JPL
Page G13. Reprinted with permission of MIT Lincoln Laboratory, Lexington, Massachusetts
Page G14. JPL
Page G15. *Top:* Reprinted with permission of MIT Lincoln Laboratory, Lexington, Massachusetts. *Bottom:* JPL
Page G16. NASA, JPL, and University of Texas's Center for Space Research / GeoForschungsZentrum Potsdam
Page G17. NASA and JPL
Page G18. JPL
Page G19. NASA, JPL, and NIMA
Page G20. NASA, GSFC, METI, ERSDAC, JAROS, and U.S./Japan ASTER Science Team
Page G21. NASA, GSFC, LaRC, JPL, and MISR Team
Page G22. NASA, JPL, and Malin Space Science Page
Page G23. JPL
Page G24. NASA and JPL
Page G25. NASA
Page G26. NASA
Page G27. NASA, The Hubble Heritage Team, StScI, and AURA
Page G28. 2MASS
Page G29. NASA, A. Fruchter, ESO Team, and StScI
Page G30. *Top:* NASA, Massimo Stiavelli, and StScI. *Bottom:* NASA, J. J. Hester, Arizona State University
Page G31. NASA and JPL-Caltech
Page G32. JPL